T0205843

LONDON MATHEMATICAL SOCIETY LECTURE NOTE SERIES

The titles below are available from booksellers, or from Cambridge University Press at
http://www.cambridge.org/mathematics

339 Groups St Andrews 2005 I, C.M. CAMPBELL, M.R. QUICK, E.F. ROBERTSON & G.C. SMITH (eds)
340 Groups St Andrews 2005 II, C.M. CAMPBELL, M.R. QUICK, E.F. ROBERTSON & G.C. SMITH (eds)
341 Ranks of elliptic curves and random matrix theory, J.B. CONREY, D.W. FARMER, F. MEZZADRI & N.C. SNAITH (eds)
342 Elliptic cohomology, H.R. MILLER & D.C. RAVENEL (eds)
343 Algebraic cycles and motives I, J. NAGEL & C. PETERS (eds)
344 Algebraic cycles and motives II, J. NAGEL & C. PETERS (eds)
345 Algebraic and analytic geometry, A. NEEMAN
346 Surveys in combinatorics 2007, A. HILTON & J. TALBOT (eds)
347 Surveys in contemporary mathematics, N. YOUNG & Y. CHOI (eds)
348 Transcendental dynamics and complex analysis, P.J. RIPPON & G.M. STALLARD (eds)
349 Model theory with applications to algebra and analysis I, Z. CHATZIDAKIS, D. MACPHERSON, A. PILLAY & A. WILKIE (eds)
350 Model theory with applications to algebra and analysis II, Z. CHATZIDAKIS, D. MACPHERSON, A. PILLAY & A. WILKIE (eds)
351 Finite von Neumann algebras and masas, A.M. SINCLAIR & R.R. SMITH
352 Number theory and polynomials, J. MCKEE & C. SMYTH (eds)
353 Trends in stochastic analysis, J. BLATH, P. MÖRTERS & M. SCHEUTZOW (eds)
354 Groups and analysis, K. TENT (ed)
355 Non-equilibrium statistical mechanics and turbulence, J. CARDY, G. FALKOVICH & K. GAWEDZKI
356 Elliptic curves and big Galois representations, D. DELBOURGO
357 Algebraic theory of differential equations, M.A.H. MACCALLUM & A.V. MIKHAILOV (eds)
358 Geometric and cohomological methods in group theory, M.R. BRIDSON, P.H. KROPHOLLER & I.J. LEARY (eds)
359 Moduli spaces and vector bundles, L. BRAMBILA-PAZ, S.B. BRADLOW, O. GARCÍA-PRADA & S. RAMANAN (eds)
360 Zariski geometries, B. ZILBER
361 Words: Notes on verbal width in groups, D. SEGAL
362 Differential tensor algebras and their module categories, R. BAUTISTA, L. SALMERÓN & R. ZUAZUA
363 Foundations of computational mathematics, Hong Kong 2008, F. CUCKER, A. PINKUS & M.J. TODD (eds)
364 Partial differential equations and fluid mechanics, J.C. ROBINSON & J.L. RODRIGO (eds)
365 Surveys in combinatorics 2009, S. HUCZYNSKA, J.D. MITCHELL & C.M. RONEY-DOUGAL (eds)
366 Highly oscillatory problems, B. ENGQUIST, A. FOKAS, E. HAIRER & A. ISERLES (eds)
367 Random matrices: High dimensional phenomena, G. BLOWER
368 Geometry of Riemann surfaces, F.P. GARDINER, G. GONZÁLEZ-DIEZ & C. KOUROUNIOTIS (eds)
369 Epidemics and rumours in complex networks, M. DRAIEF & L. MASSOULIÉ
370 Theory of p-adic distributions, S. ALBEVERIO, A.YU. KHRENNIKOV & V.M. SHELKOVICH
371 Conformal fractals, F. PRZYTYCKI & M. URBAŃSKI
372 Moonshine: The first quarter century and beyond, J. LEPOWSKY, J. MCKAY & M.P. TUITE (eds)
373 Smoothness, regularity and complete intersection, J. MAJADAS & A. G. RODICIO
374 Geometric analysis of hyperbolic differential equations: An introduction, S. ALINHAC
375 Triangulated categories, T. HOLM, P. JØRGENSEN & R. ROUQUIER (eds)
376 Permutation patterns, S. LINTON, N. RUŠKUC & V. VATTER (eds)
377 An introduction to Galois cohomology and its applications, G. BERHUY
378 Probability and mathematical genetics, N. H. BINGHAM & C. M. GOLDIE (eds)
379 Finite and algorithmic model theory, J. ESPARZA, C. MICHAUX & C. STEINHORN (eds)
380 Real and complex singularities, M. MANOEL, M.C. ROMERO FUSTER & C.T.C WALL (eds)
381 Symmetries and integrability of difference equations, D. LEVI, P. OLVER, Z. THOMOVA & P. WINTERNITZ (eds)
382 Forcing with random variables and proof complexity, J. KRAJÍČEK
383 Motivic integration and its interactions with model theory and non-Archimedean geometry I, R. CLUCKERS, J. NICAISE & J. SEBAG (eds)
384 Motivic integration and its interactions with model theory and non-Archimedean geometry II, R. CLUCKERS, J. NICAISE & J. SEBAG (eds)
385 Entropy of hidden Markov processes and connections to dynamical systems, B. MARCUS, K. PETERSEN & T. WEISSMAN (eds)
386 Independence-friendly logic, A.L. MANN, G. SANDU & M. SEVENSTER
387 Groups St Andrews 2009 in Bath I, C.M. CAMPBELL *et al* (eds)
388 Groups St Andrews 2009 in Bath II, C.M. CAMPBELL *et al* (eds)
389 Random fields on the sphere, D. MARINUCCI & G. PECCATI

390 Localization in periodic potentials, D.E. PELINOVSKY
391 Fusion systems in algebra and topology, M. ASCHBACHER, R. KESSAR & B. OLIVER
392 Surveys in combinatorics 2011, R. CHAPMAN (ed)
393 Non-abelian fundamental groups and Iwasawa theory, J. COATES *et al* (eds)
394 Variational problems in differential geometry, R. BIELAWSKI, K. HOUSTON & M. SPEIGHT (eds)
395 How groups grow, A. MANN
396 Arithmetic differential operators over the p-adic integers, C.C. RALPH & S.R. SIMANCA
397 Hyperbolic geometry and applications in quantum chaos and cosmology, J. BOLTE & F. STEINER (eds)
398 Mathematical models in contact mechanics, M. SOFONEA & A. MATEI
399 Circuit double cover of graphs, C.-Q. ZHANG
400 Dense sphere packings: a blueprint for formal proofs, T. HALES
401 A double Hall algebra approach to affine quantum Schur-Weyl theory, B. DENG, J. DU & Q. FU
402 Mathematical aspects of fluid mechanics, J.C. ROBINSON, J.L. RODRIGO & W. SADOWSKI (eds)
403 Foundations of computational mathematics, Budapest 2011, F. CUCKER, T. KRICK, A. PINKUS & A. SZANTO (eds)
404 Operator methods for boundary value problems, S. HASSI, H.S.V. DE SNOO & F.H. SZAFRANIEC (eds)
405 Torsors, étale homotopy and applications to rational points, A.N. SKOROBOGATOV (ed)
406 Appalachian set theory, J. CUMMINGS & E. SCHIMMERLING (eds)
407 The maximal subgroups of the low-dimensional finite classical groups, J.N. BRAY, D.F. HOLT & C.M. RONEY-DOUGAL
408 Complexity science: the Warwick master's course, R. BALL, V. KOLOKOLTSOV & R.S. MACKAY (eds)
409 Surveys in combinatorics 2013, S.R. BLACKBURN, S. GERKE & M. WILDON (eds)
410 Representation theory and harmonic analysis of wreath products of finite groups, T. CECCHERINI-SILBERSTEIN, F. SCARABOTTI & F. TOLLI
411 Moduli spaces, L. BRAMBILA-PAZ, O. GARCÍA-PRADA, P. NEWSTEAD & R.P. THOMAS (eds)
412 Automorphisms and equivalence relations in topological dynamics, D.B. ELLIS & R. ELLIS
413 Optimal transportation, Y. OLLIVIER, H. PAJOT & C. VILLANI (eds)
414 Automorphic forms and Galois representations I, F. DIAMOND, P.L. KASSAEI & M. KIM (eds)
415 Automorphic forms and Galois representations II, F. DIAMOND, P.L. KASSAEI & M. KIM (eds)
416 Reversibility in dynamics and group theory, A.G. O'FARRELL & I. SHORT
417 Recent advances in algebraic geometry, C.D. HACON, M. MUSTAŢĂ & M. POPA (eds)
418 The Bloch-Kato conjecture for the Riemann zeta function, J. COATES, A. RAGHURAM, A. SAIKIA & R. SUJATHA (eds)
419 The Cauchy problem for non-Lipschitz semi-linear parabolic partial differential equations, J.C. MEYER & D.J. NEEDHAM
420 Arithmetic and geometry, L. DIEULEFAIT *et al* (eds)
421 O-minimality and Diophantine geometry, G.O. JONES & A.J. WILKIE (eds)
422 Groups St Andrews 2013, C.M. CAMPBELL *et al* (eds)
423 Inequalities for graph eigenvalues, Z. STANIĆ
424 Surveys in combinatorics 2015, A. CZUMAJ *et al* (eds)
425 Geometry, topology and dynamics in negative curvature, C.S. ARAVINDA, F.T. FARRELL & J.-F. LAFONT (eds)
426 Lectures on the theory of water waves, T. BRIDGES, M. GROVES & D. NICHOLLS (eds)
427 Recent advances in Hodge theory, M. KERR & G. PEARLSTEIN (eds)
428 Geometry in a Fréchet context, C. T. J. DODSON, G. GALANIS & E. VASSILIOU
429 Sheaves and functions modulo p, L. TAELMAN
430 Recent progress in the theory of the Euler and Navier-Stokes equations, J.C. ROBINSON, J.L. RODRIGO, W. SADOWSKI & A. VIDAL-LÓPEZ (eds)
431 Harmonic and subharmonic function theory on the real hyperbolic ball, M. STOLL
432 Topics in graph automorphisms and reconstruction (2nd Edition), J. LAURI & R. SCAPELLATO
433 Regular and irregular holonomic D-modules, M. KASHIWARA & P. SCHAPIRA
434 Analytic semigroups and semilinear initial boundary value problems (2nd Edition), K. TAIRA
435 Graded rings and graded Grothendieck groups, R. HAZRAT
436 Groups, graphs and random walks, T. CECCHERINI-SILBERSTEIN, M. SALVATORI & E. SAVA-HUSS (eds)
437 Dynamics and analytic number theory, D. BADZIAHIN, A. GORODNIK & N. PEYERIMHOFF (eds)
438 Random walks and heat kernels on graphs, M.T. BARLOW
439 Evolution equations, K. AMMARI & S. GERBI (eds)
440 Surveys in combinatorics 2017, A. CLAESSON *et al* (eds)
441 Polynomials and the mod 2 Steenrod algebra I, G. WALKER & R.M.W. WOOD
442 Polynomials and the mod 2 Steenrod algebra II, G. WALKER & R.M.W. WOOD
443 Asymptotic analysis in general relativity, T. DAUDÉ, D. HÄFNER & J.-P. NICOLAS (eds)
444 Geometric and cohomological group theory, P.H. KROPHOLLER, I.J. LEARY, C. MARTÍNEZ-PÉREZ & B.E.A. NUCINKIS (eds)
445 Introduction to hidden semi-Markov models, J. VAN DER HOEK & R.J. ELLIOTT
446 Advances in two-dimensional homotopy and combinatorial group theory, W. METZLER & S. ROSEBROCK (eds)
447 New directions in locally compact groups, P.-E. CAPRACE & N. MONOD (eds)
448 Synthetic differential topology, M.C. BUNGE, F. GAGO & A.M. SAN LUIS

London Mathematical Society Lecture Note Series: 448

Synthetic Differential Topology

MARTA BUNGE
McGill University, Montréal

FELIPE GAGO
Universidade de Santiago de Compostela, Spain

ANA MARÍA SAN LUIS
Universidad de Oviedo, Spain

CAMBRIDGE
UNIVERSITY PRESS

University Printing House, Cambridge CB2 8BS, United Kingdom

One Liberty Plaza, 20th Floor, New York, NY 10006, USA

477 Williamstown Road, Port Melbourne, VIC 3207, Australia

314-321, 3rd Floor, Plot 3, Splendor Forum, Jasola District Centre, New Delhi - 110025, India

79 Anson Road, #06-04/06, Singapore 079906

Cambridge University Press is part of the University of Cambridge.

It furthers the University's mission by disseminating knowledge in the pursuit of education, learning and research at the highest international levels of excellence.

www.cambridge.org
Information on this title: www.cambridge.org/9781108447232
DOI: 10.1017/9781108553490

First published 2018

A catalogue record for this publication is available from the British Library

Library of Congress Cataloging in Publication data
Names: Bunge, M. (Marta), author. | Gago Couso, Felipe, author. |
San Luis Fernandez, Ana Maria, author.
Title: Synthetic differential topology / Marta Bunge (McGill University, Montreal),
Felipe Gago (Universidade de Santiago de Compostela, Spain),
Ana Maria San Luis (Universidad de Oviedo, Spain).
Description: Cambridge : Cambridge University Press, 2018. |
Series: London Mathematical Society lecture note series ; 448 |
Includes bibliographical references and index.
Identifiers: LCCN 2017053760 | ISBN 9781108447232 (hardback : alk. paper)
Subjects: LCSH: Differential topology. | Geometry, Differential. | Mappings (Mathematics)
Classification: LCC QA613.6.B86 2018 | DDC 514/.72–dc23
LC record available at https://lccn.loc.gov/2017053760

ISBN 978-1-108-44723-2 Paperback

Contents

Preface

The subject of synthetic differential geometry has its origins in lectures and papers by F. William Lawvere, most notably [72], but see also [74, 76]. It extends the pioneering work of Charles Ehresmann [40] and André Weil [111] to the setting of a topos [73, 55]. It is synthetic (as opposed to analytic) in that the basic concepts of the differential calculus are introduced by axioms rather than by definition using limits or other quantitative data. It attempts to capture the classical concepts of differential geometry in an intuitive fashion using the rich structure of a topos (finite limits, exponentiation, subobject classifier) in order to conceptually simplify both the statements and their proofs. The fact that the intrinsic logic of any topos model of the theory is necessarily Heyting (or intuitionistic) rather than Boolean (or classical) plays a crucial role in its development. It is well adapted to the study of classical differential geometry by virtue of some of its models.

This book is intended as a natural extension of synthetic differential geometry (SDG), in particular of the book by Anders Kock [61] to (a subject that we here call) synthetic differential topology (SDT). Whereas the basic axioms of SDG are the representability of jets (of smooth mappings) by tiny objects of an algebraic nature, those of SDT are the representability of germs (of smooth mappings) by tiny objects of a logical sort introduced by Jacques Penon [96, 94, 95]. In both cases, additional axioms and postulates are added to the basic ones in order to develop special portions of the theory.

In a first part we include those portions of topos theory and of synthetic differential geometry that should minimally suffice for a reading of the book. As an illustration of the benefits of working synthetically within topos theory we include in a second part a version of the theory of connections and sprays [28, 22] as well as one of the calculus of variations [52, 27]. The basic axioms for SDT were introduced in [20, 25, 26] and are the contents of the third part of this book. The full force of SDT is employed in the fourth part of the book and consists of an application to the theory of stable germs of smooth mappings including Mather's theorem [20, 26, 103] and Morse theory on the classification of singularities [44, 45, 46]. The fifth part of the book recalls the notion of a well adapted model of SDG in the sense of [32, 10] and extends it to one of SDT. In this same part, and under the assumption of the existence of a well adapted model of SDT, a theory of unfoldings is given as a particular case of the general theory, unlike what is done in the classical case [110]. The sixth part of the book is devoted to exhibiting one such well adapted model of SDT, namely a Grothendieck topos $\mathscr{G}$ constructed by Eduardo Dubuc [34] using the algebraic theory [70] of C^∞-rings [72] and germ determined (or local) ideals. On account of the existence of a well adapted model of SDT, several classical results can be recovered. In these applications of SDG and SDT to classical mathematics, it should be noted that not only do they profit from the rich structure of a topos, not available when working in the category of smooth manifolds, but also that the results so obtained are often of a greater generality and conceptual simplicity than their classical counterparts.

Acknowledgements

We are grateful to F. William Lawvere and Andrée Ehresmann for their valuable input and constant support in matters related to the subject of this book. We are also grateful to Anders Kock for his helpful questions and comments on some portions of an earlier version of the book. Useful remarks from George Janelidze and Thomas Streicher are also gratefully acknowledged.

Introduction

This book deals with a subject that extends synthetic differential geometry [61] to differential topology, in particular to the theory of smooth mappings and their singularities. The setting is that of category theory [86] in general and of topos theory [55] in particular. An excellent introduction to both subjects including applications to several topics (among them synthetic differential geometry) is [79]. The subject of toposes in logic and logic in toposes is illustrated in [21], an article intended for philosophers. Our book is intended as the basis for an advanced course or seminar whose only prerequisite is a reasonable acquaintance with category theory, logic, commutative algebra, infinitesimal calculus, general topology, differential geometry and topology.

Motivated by the desire to employ category theory in a non-trivial way in (elementary) Physics, Lawvere [72] in 1967 gave lectures on 'Categorical Dynamics' which would turn out to be the beginning of a new subject, a branch of (applied) category theory which came to be labelled 'synthetic differential geometry' (SDG), as opposed to 'analytic' which relies heavily on the use of coordinates. What Lawvere proposed was to do Dynamics, not in the context of manifolds, but in a category $\mathscr{E}$, different from the category $\mathscr{M}^\infty$ of smooth paracompact manifolds in several respects : (1) in $\mathscr{E}$, 'the line' would be represented by an object R which, unlike the classical reals, would not be a field but just a commutative ring in which nilpotent elements could be thought of as infinitesimals, and (2) in $\mathscr{E}$, unlike in $\mathscr{M}^\infty$, all finite limits and exponentials would be assumed to exist so that, for the objects of $\mathscr{E}$ thought of as 'smooth spaces' and for the morphisms of $\mathscr{E}$ thought of as 'smooth maps', one could form all fibred products (not just the transversal ones) and something so basic as the smooth space of all smooth maps between two smooth spaces would exist.

The idea of introducing infinitesimals so as to render more intuitive the foundations of analysis was not new. On the one hand, there are non-standard models of analysis [102] in which the non-standard reals have infinitesimals, but where the field property is retained and so the possibility of dividing by non-zero elements gives infinitely large non-zero reals. On the other hand, commutative algebra deals with nilpotent elements in rings and treats them as infinitesimals of some kind. However, on account of the assumptions made about the line, SDG is quite different from non-standard analysis and goes beyond commutative algebra as it has models arising also from differential geometry and analysis and not just from commutative algebra and algebraic geometry.

It is customary to assume further that $\mathscr{E}$ is a topos, even a Grothendieck topos [4], although the Grothendieck toposes that are usually considered as models of SDG are C^∞ versions of those devised by Grothendieck to do algebraic geometry. The idea of working in a topos is not new either as Chen, in 1977, constructed a 'gros' topos for the same purpose, but one in which there was no room for infinitesimals [29]. The two conditions imposed on $\mathscr{E}$ by Lawvere were put to work in SDG by means of the basic axiom of the theory, namely, the axiom that states that R be 'of line type', also known as the 'Kock-Lawvere axiom', and which we discuss in the first part of this book. As stated already, these developments owe much to the lead of André Weil [111] and Charles Ehresmann [40], although the SDG treatment of classical differential geometry differs from those in that the basic constructions in SDG are more natural than in theirs; for instance, tangent spaces are representable as some sort of function spaces, whereas this is not the case in the approach by means of 'near points'.

Although the origins of SDG were strongly influenced by several developments, such as Robinson's non-standard analysis, Weil and Ehresmann's theory of infinitely near points, Grothendieck's use of toposes in algebraic geometry, and Chen's gros toposes in his treatment of the calculus of variations, it differs from all four of them. It differs from non-standard analysis in that SDG is carried out in a topos whose internal logic is necessarily non-Boolean and where R is not a field. It differs from the Weil and Ehresmann's treatment in that the tangent spaces and other spaces of jets are presented as function spaces which need no special construction as they exist naturally by virtue of the topos axioms. It differs from Chen's gros topos models in that in SDG infinitesimals exist and so permit intuitive and direct arguments in the style of non-standard analysis. It differs from the use of Grothendieck toposes in algebraic geometry in that the well adapted models for SDG, by which it is meant models with $\mathscr{E}$ a topos and R a ring of line type in the generalized sense, for which a full embedding $\mathscr{M}^\infty \hookrightarrow \mathscr{E}$ of the category of smooth manifolds exists and has some good properties, such as sending $\mathbb{R}$ to R, preserving limits that exist and constructions that are available, are quite different although in a sense analogous to those arising from the affine schemes in that the smooth aspect and corresponding notion of C^∞-ring is the basis for constructing such models [32, 34, 33].

The introduction of the intrinsic (or Penon) topology [96, 94] on any object of a topos $\mathscr{E}$ and, for a model $(\mathscr{E}, R)$ of SDG, that of the object $\Delta(n) = \neg\neg\{0\} \hookrightarrow R^n$ of 'all infinitesimals' in R^n, intended to represent germs at $0 \in R^n$ of smooth mappings from R^n to R, opened up the way to synthetic differential topology (SDT). In particular, a synthetic theory of stable mappings to be based on SDT was proposed as a theory which extended SDG by means of axioms and postulates (germs representability, tinyness of the representing objects, infinitesimal inversion, infinitesimal integration of vector fields, density of regular values) introduced formally in [20, 24]. The main application of Mather's theorem (infinitesimally stable germs are stable) is a useful tool for the classification of stable mappings. We give two proofs of it here, one which (as in the classical case) makes use of a 'Weierstrass preparation theorem' [26], and another [103] which does not. As in the classical case, the notion of a generic property was introduced in SDT [44] and was shown therein to be satisfied by the stable germs. In the case of Morse germs [46] in SDT, genericity is shown to follow from the facts that Morse germs are both stable and dense.

A general way to proceed in applying SDG or SDT to classical differential geometry or topology is as follows. First, a classical problem or statement is formulated in the internal language of the topos $\mathscr{E}$, where $(\mathscr{E}\ R)$ is a well adapted model of the synthetic theory **T** to be used (for instance SDT or just SDG), in such a way that when applying the global sections functor $\Gamma = \mathrm{Hom}(1, -) : \mathscr{E} \longrightarrow \mathit{Set}$ the original problem or statement be recovered. The second step consists in making use of the rich structure of the topos $\mathscr{E}$ (finite limits, exponentiation, infinitesimals) in order to give definitions or prove theorems in a conceptually simpler and more intuitive fashion than in their classical forms. It is often the case that this step requires an enrichment of the synthetic theory **T** through the adoption of additional axioms. A guideline for the selection of such axioms is restricted by the need to ultimately prove their consistency with the axioms of **T**. This requirement renders the subject less trivial than what it may appear at first, as the axioms should also be as few and as basic as possible. The verification of the validity of the additional axioms in $\mathscr{E}$ constitutes the third step. The fourth and final step is to reinterpret the internal solution to the problem as a classical statement, either by applying the global sections functor Γ (which, however, has poor preservation properties in general) or by restricting the objects involved to those that arise from a classical set-up via the embedding $\iota : \mathscr{M}^\infty \hookrightarrow \mathscr{E}$.

The applications of SDG to classical differential geometry and topology that are given in this book are to the theory of connections and sprays [28], the calculus of variations [27], the stability theory of smooth mappings [26, 44, 103], and Morse theory [45]. In order to carry out such applications, an acquaintance with the appropriate portions of the subject matter itself is naturally a prerequisite. There are several references where the classical theories of connections and the calculus of variations are expounded. For the former, our sources were [1], [93], and [98]. For the latter, we used [9] and [49]. Among the classical sources for the theory of smooth manifolds and their singularities including Morse theory, our sources were [3], [11], [13], [47], [48], [51], [54], [68], [82], [83, 84, 85], [87], [91], [104], [106], [108], and [110]. In this case, what is needed in order to derive the classical theorems (and generalizations of them) from their versions within SDT is to establish the existence of a well adapted model of the latter. This is precisely what concerns the last part of the book.

This book consists of six parts. In the *first part* we review all basic notions of the theory of toposes that are needed in the sequel. Of particular importance are two such notions that arose in connection with applications of toposes in set theory, algebraic geometry and differential geometry and topology, to wit, atoms and Penon opens. If desired, this material could be extended to cover some of the topics from [55, 6, 12] and references therein. This is followed by a summary of the main aspects of synthetic differential geometry, which we refer to as SDG [61]. The first axiom of SDG postulates, for a topos $\mathscr{E}$ with a natural numbers object N and a commutative ring R in it, the representability of jets of mappings as mappings themselves. As a second axiom we postulate that the jet representing objects be in some sense infinitesimal. To these two axioms we add several postulates that are used in order to develop part of the differential calculus. In order to illustrate the uses of SDG for differential geometry and analysis we give, in the *second part* of this book, two different applications of it: a theory of connections and sprays [28, 22], and a version of the calculus of variations [52, 27]. In the theory of connections and sprays within SDG it is emphasized that, unlike the classical theory, the passage from connections to (geodesic) sprays need not involve integration except in infinitesimal form. In the case of the calculus of variations within SDG, it is shown that one can develop it without variations except for those in an infinitesimal guise. In both illustrations, the domain of application is the class of infinitesimally linear objects, which includes R and is closed under finite limits, exponentiation and étale descent. In particular, and in both cases, the domain of applications of SDG extends beyond their classical counterparts.

In the *third part* of this book we introduce the subject matter of the title. The origin of synthetic differential topology, which we refer to as SDT, can be traced back to the introduction [96] of an intrinsic topological structure on any object of a topos ('Penon opens'). This in turn motivated the introduction and study of general topological structures in toposes [25] and is included here as a preliminary to the specific topological structures of interest in this book, that is, the Euclidean and the weak topological structures. By synthetic differential topology (SDT) we shall understand an extension of synthetic differential geometry (SDG) obtained by adding to it axioms of a local nature—to wit, germ representability and the tinyness of the representing objects [96, 25], which are logical, rather than algebraic infinitesimals. To those, we add four postulates.

The problem of classifying all germs of smooth mappings according to their singularities is intractable. Topologists reduce the question to the consideration of stable (germs of) smooth mappings. In the context of synthetic differential topology, the entire subject is considerably simplified by the force of the axiom of the representability of germs of smooth mappings by means of logical infinitesimals. A smooth mapping is said to be stable if any infinitesimal deformation of it is equivalent to it, in the sense that under a small deformation there is no change in the nature of the function. A class of mappings is said to be generic if the class is closed under equivalence and is dense in that of all smooth mappings equipped with the Whitney topology. The main tool in the classification problem is Mather's theorem [84]. A theory of germs of smooth mappings within SDT has been developed by the authors of the present monograph [20, 44, 26, 103] and constitutes the *fourth part* of this book. The notion of stability for mappings, or for germs, is important for several reasons, one of which is its intended application in the natural sciences, as promoted by R. Thom [107]. Another reason for concentrating on stability is the simplification that it brings about to the classification of singularities. In this same part we apply the results obtained in order to give a version [45, 46] of Morse theory [87] within SDT.

In the *fifth part* of the book we introduce a notion of well adapted model of SDT, based on a previous notion of that of a well adapted model of SDG. In the *sixth part* of the book we focus on a particular model of SDT that is shown to be well adapted to the applications to classical mathematics in the sense of [32, 10]. This model is the Dubuc topos $\mathcal{G}$ [34], constructed using the notion of a C^∞-ring which is due to F.W. Lawvere and goes back to [70]. What makes this topos a well adapted model of SDT (in fact, the only one that is known, at present) is the nature of the ideals, which are germ determined or local. Some of the axioms involved in the synthetic theory for differential topology are intrinsically related to this particular model, whereas others were suggested by their potential applications to a theory of smooth mappings and their singularities. The existence of a well adapted model of SDT is what renders it relevant to classical mathematics.

PART ONE

TOPOSES AND DIFFERENTIAL GEOMETRY

With the introduction of toposes as a categorical surrogate of set theory whose underlying logic is Heyting and where no appeal to an axiom of choice is permitted, it became possible to formally introduce infinitesimals in the study of differential geometry in the same spirit as that of the work carried out by Charles Ehresmann and André Weil in the 50s. Whereas the known models that are well adapted for the applications to classical mathematics are necessarily Grothendieck toposes, the theory of such is not enough to express the richness of the synthetic theory. For the latter one needs the internal language that is part of the theory of toposes and that is based on the axiom of the existence of a subobjects classifier. This first part is an introduction to topos theory and to synthetic differential geometry, both of which originated in the work of F.W. Lawvere. These introductory presentations will gradually be enriched as we proceed, but only as needed for the purposes of this book.

1
Topos Theory

A topos may be viewed as a universe of variable sets, as a generalized topological space and as a semantical universe for higher-order logic. However, neither of these views is enough in itself to describe, even informally, what a topos is or what it is good for. Topos theory permits a conceptually richer view of classical mathematics and it is in that capacity that its strength lies. The notion of a topos is due to F. W. Lawvere and M. Tierney [73]. It stemmed from an original blending of previously unrelated areas, to wit, the theory of schemes due to A. Grothendieck [4] and constructivist mathematics [39, 53, 109]. An account of the developments of topos theory since its inception is given in [77]. This chapter gives an introduction to just those parts of topos theory on which this book is based. To this end we assume some familiarity with category theory which can be acquired from [86]. Our standard reference is [55], but see also [6]. An excellent source that deals with some of the applications of topos theory, including synthetic differential geometry, is [79].

1.1 Basic Notions of Toposes

Definition 1.1 A topos is a category $\mathscr{E}$ such that the following axioms are satisfied:

(i) (**Finite limits**) There is a terminal object 1 in $\mathscr{E}$ and any pair of morphisms $f: A \longrightarrow C$ and $g: B \longrightarrow C$ has a pullback

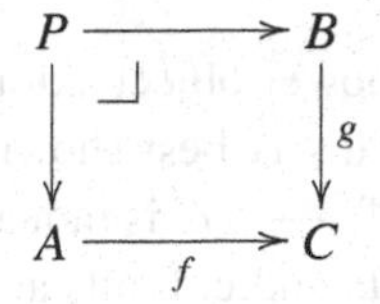

(ii) **(Power objects)** For each object A there is an object $\mathscr{P}(A)$ and a relation $\in_A$ from A to $\mathscr{P}(A)$, i.e., a monomorphism $\in_A \rightarrowtail A \times \mathscr{P}(A)$, which is generic in the sense that, for any relation $R \rightarrowtail A \times B$, there is a unique morphism $r\colon B \to \mathscr{P}(A)$ such that the diagram

$$\begin{array}{ccc} R & \longrightarrow & \in_A \\ \downarrow & \lrcorner & \downarrow \\ A \times B & \xrightarrow[\mathrm{id}_A \times r]{} & A \times \mathscr{P}(A) \end{array}$$

is a pullback.

Definition 1.2 (Peano-Lawvere axiom) A topos $\mathscr{E}$ is said to have a *natural numbers object* if there is an object N of $\mathscr{E}$, together with morphisms

$$0\colon 1 \longrightarrow N, \quad s\colon N \longrightarrow N$$

such that for any object X and any pair of morphisms

$$x\colon 1 \longrightarrow X, \quad u\colon X \longrightarrow X$$

in $\mathscr{E}$, there exists a unique morphism $f\colon N \longrightarrow X$ in $\mathscr{E}$ such that the diagram

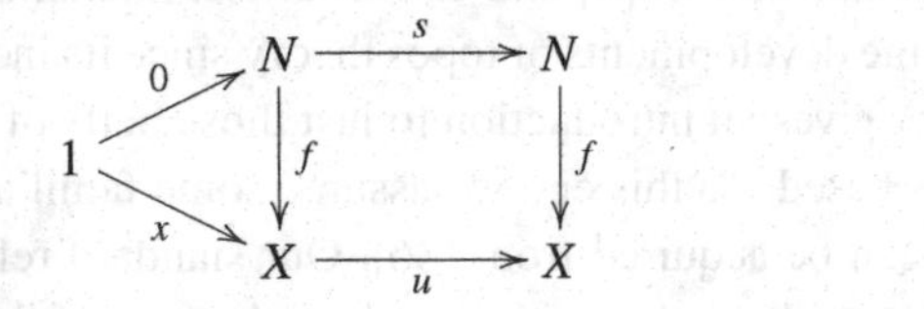

commutes. (All the usual rules of arithmetic follow from this definition.)

The object $\Omega = \mathscr{P}(1)$ in a topos $\mathscr{E}$ comes with a morphism $\top\colon 1 \longrightarrow \Omega$ (for 'true'), together with which it becomes a *subobjects classifier* in the sense that for an object A, every subobject $u\colon A' \rightarrowtail A$ has a unique 'characteristic function' $\varphi\colon A \longrightarrow \Omega$, so that u is the pullback of $\top$ along φ.

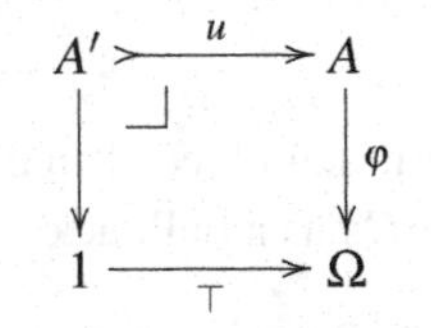

This is a consequence of the power object axiom. Another consequence is the existence of finite colimits. This is best shown by observing that the power object functor $\mathscr{P} = \Omega^{(-)}\colon \mathscr{E}^{\mathrm{op}} \longrightarrow \mathscr{E}$ is monadic (Paré's theorem [92]).

The existence of finite limits and colimits and the universal property of the pair $(\Omega, \top\colon 1 \to \Omega)$ make it possible to define new morphisms corresponding to logical connectives, to wit, $\bot\colon 1 \to \Omega$ (for 'false'), $\wedge\colon \Omega\times\Omega \to \Omega$ ('conjunction') , $\vee\colon \Omega\times\Omega \to \Omega$ ('disjunction'), $\Rightarrow\colon \Omega\times\Omega \longrightarrow \Omega$ ('implication'), and $\neg\colon \Omega \longrightarrow \Omega$ ('negation'). With these, Ω becomes a *Heyting algebra*.

It also follows from the axioms for a topos that *exponentiation* exists. This property which, together with the existence of finite limits and the existence of a subobjects classifier, provides for an alternative definition of the notion of a topos, says the following: Given any two objects A, B of $\mathscr{E}$, there is given an object B^A and a morphism $\mathrm{ev}_{A,B}\colon A\times B^A \longrightarrow B$ satisfying a universal property which can be summarized by saying that for each object A, the functor $A\times(-)\colon \mathscr{E} \longrightarrow \mathscr{E}$ has a right adjoint

$$A\times(-) \dashv (-)^A\colon \mathscr{E} \longrightarrow \mathscr{E}\,.$$

Equivalently, for each object C, there is a bijection between morphisms $C \longrightarrow B^A$ and morphisms $C\times A \longrightarrow B$, natural in all three variables. The way to construct exponentials in a topos is to mimic the set-theoretical construction: a function from A to B is a relation from A to B that is single-valued and everywhere defined.

We can transfer to $\mathscr{P}(A) = \Omega^A$ the Heyting algebra structure from Ω and moreover, for any map $f\colon A \to B$ in $\mathscr{E}$ the map $f^{-1}\colon \mathscr{P}(B) \to \mathscr{P}(A)$ induced by pulling back along f, when viewed as a functor between the categories $\mathscr{P}(B)$ and $\mathscr{P}(A)$ in the usual way, has both a left and a right adjoint

$$\exists_f \dashv f^{-1} \dashv \forall_f,$$

which enable the internalization of quantifiers.

Remark 1.3 On account of the fact that the subobjects classifier Ω (and more generally any power object $\mathscr{P}(A)$) in a topos $\mathscr{E}$ becomes naturally a Heyting algebra, we may state that the logic of a topos is intuitionistic. We shall avoid doing so since it may lead to misunderstandings about the nature of topos theory while denying Heyting his due credit. We shall therefore use the expression 'Heyting logic' except when quoting authors who use 'intuitionistic logic' instead. As for classical logic, we shall mostly use the terminology 'Boolean logic' for it.

More generally, it can be shown that, for $\mathscr{E}$ a topos, any first-order formula $\varphi(x_1,\ldots,x_n)$, with free variables of sorts objects $A_1,\ldots,A_n$ of $\mathscr{E}$, admits an *extension* in the form of a subobject

$$[[(x_1,\ldots,x_n) \mid \varphi(x_1,\ldots,x_n)]] \rightarrowtail A_1\times\cdots\times A_n\,.$$

The so-called *Kripke-Joyal semantics* [12, §6.6] consists of the rules governing these interpretations and extends the semantics of S. Kripke [67] for intuitionistic (or Heyting) logic.

Any topos $\mathscr{E}$ has *stable image factorizations* for its morphisms. This means that, given any morphism $f\colon A \longrightarrow B$ in $\mathscr{E}$, there exists a factorization

$$A \twoheadrightarrow I_f \rightarrowtail B$$

into an epimorphism followed by a monomorphism, which is universal among all such factorizations. The monomorphism $I_f \rightarrowtail B$ is said to be the *image* of f. Furthermore, image factorizations are preserved by pullback functors.

The terminology *surjective* is sometimes used in this book. A formal definition follows.

Definition 1.4 A morphism $f\colon A \longrightarrow B$ in a topos $\mathscr{E}$ is said to be *surjective* if in the image factorization of f the monomorphism $I_f \rightarrowtail B$ is an isomorphism.

Let $f\colon A \longrightarrow B$ be a morphism in a topos $\mathscr{E}$. Then f is surjective if and only if it is an epimorphism.

Example 1.5 A first example of a topos is the category *Set* of sets and functions, a model of ZFC, or Zermelo-Fraenkel Set Theory with Choice. It is well known that it has all finite limits. For a set A, $\mathscr{P}(A)$ is its power set and $\in_A$ is the membership relation. As a universe of variable sets, its objects admit no genuine variation, or are constant. All of higher-order logic is interpretable in *Set*. As a generalized topological space it is trivial, i.e., the space has only one point.

Example 1.6 A second example of a topos is any category $\mathrm{Sh}(X)$ of sheaves on a topological space X, which we proceed to describe in two different ways. For X a topological space, a *sheaf* on X is a pair (E,π) where E is a topological space and

$$\pi\colon E \longrightarrow X$$

is a local homeomorphism, which means a continuous map such that each point $e \in E$ has an open neighbourhood which is mapped homeomorphically onto an open neighborhood of $\pi(x)$. A *sheaf morphism* from $\pi\colon E \longrightarrow X$ to $\pi'\colon E' \longrightarrow X$ is given by a continuous map $f\colon E \longrightarrow E'$ which preserves the fibres, in other words, such that the diagram

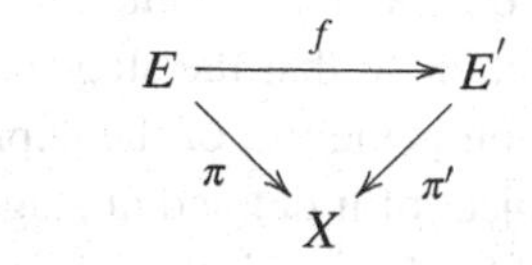

commutes. The data just given can easily be shown to be a category, which is denoted $\mathrm{Sh}(X)$.

An alternative description of the category $\mathrm{Sh}(X)$ is a direct generalization of the notion of a topological space, as follows. A *section* of a local homeomorphism $\pi\colon E \longrightarrow X$ above an open subset $U \subset X$ is a continuous map $s\colon U \longrightarrow E$ such that $\pi \circ s = \mathrm{id}_U$. Taking sections defines a functor F on the opposite of the category $\mathscr{O}(X)$ of open subsets of X and inclusions into the category *Set* of sets and functions. The rule which to an open inclusion $V \subset U$ assigns a function $F(U) \longrightarrow F(V)$ is the restriction $s \mapsto s|_V$ of a section on U to V. Let $\{U_\alpha\}$ be any open covering of U, and let s_α be any family of sections that is compatible on the intersections $U_\alpha \cap U_\beta$, in the sense that

$$s_\alpha|_{U_\alpha \cap U_\beta} = s_\beta|_{U_\alpha \cap U_\beta}.$$

Then there exists a unique section $s \in F(U)$ such that $s|_{U_\alpha} = s_\alpha$ for each α. Abstracting this property gives an alternative definition of a sheaf, to wit, any functor $F\colon \mathscr{O}(X)^{\mathrm{op}} \longrightarrow$ *Set* satisfying the above condition for any open covering of X.

A sheaf F on X may be viewed as a 'variable set' by interpreting the elements of $F(U)$ for an open $U \subset X$ as 'elements of F defined at stage U'.

We have seen that the category $\mathrm{Sh}(X)$, which is in an obvious sense a generalized topological space, is also a universe of variable sets. It was surprising for Lawvere to discover that any category $\mathrm{Sh}(X)$ and, more generally, any Grothendieck topos $\mathrm{Sh}(\mathbf{C},J)$ of *Set*-based sheaves on a site $(\mathbf{C},J)$, is a semantic universe for higher-order logic. In particular, categories of sheaves on a topological space are toposes.

Remark 1.7 In the example of the topos $\mathrm{Sh}(X)$ (and more generally that of any Grothendieck topos $\mathrm{Sh}(\mathbf{C},J)$) we have used the expression 'elements of F defined at a certain stage' for an object F, where it is clear what is meant. For an arbitrary topos $\mathscr{E}$, this same expression is often used as well (and we do so in several parts of the book) with a similar though more general meaning. Explicitly, for A and F objects of $\mathscr{E}$, we write $a \in_A F$ and say that 'a is an element of F defined at stage A' to mean that it is given as a morphism $a\colon A \longrightarrow F$ in $\mathscr{E}$. In toposes $\mathscr{E}$ of sheaves on a topological space, or more generally on a site $(\mathbf{C}, J)$, the stages A are taken to be objects of the site which in the case of $\mathrm{Sh}(X)$ are the open subsets $U \subset X$. In the general case of a topos $\mathscr{E}$ we are not able to make such a restriction so that the stages can be arbitrary objects of the topos. Another convention is the use of the expression 'global section' of an object F to mean any object of it defined at stage 1, the terminal object of $\mathscr{E}$. If we simply write $a \in F$ for F an object of $\mathscr{E}$, it means that the stage of definition is of no importance in the context in which it is used, and not necessarily that it is a global section of F. In the rest of the book we shall use these expressions in two different ways. For simplicity, we sometimes write $a \in_A F$ as a morphism $a\colon A \longrightarrow F$. It is only when some change of stage (or 'change of base') is involved that the former rather than the latter expression may be preferable.

Example 1.8 Another class of examples of toposes is that of categories of *Set*-valued functors $\mathit{Set}^{\mathbf{C}}$ for any small category $\mathbf{C}$. In particular, the Kripke models for intuitionistic (or Heyting) logic are instances of the latter, where the small categories $\mathbf{C}$ are partially ordered sets. More generally, for $\mathscr{E}$ a topos and $\mathbf{C}$ a category internal to $\mathscr{E}$, the category $\mathscr{E}^{\mathbf{C}}$ of diagrams on $\mathbf{C}$ is again a topos. A characterization of categories of diagrams on a base topos $\mathscr{E}$ among all toposes was given in [18] and generalizes (as well as internalizes) the characterization of categories of *Set*-valued functors $\mathit{Set}^{\mathbf{C}}$ among locally small categories given in [16], itself inspired by the elementary theory of the category of sets given by F. W. Lawvere [71].

Example 1.9 Grothendieck toposes [4] constitute a common generalization of both categories of the form $\mathrm{Sh}(X)$ for X a topological space and categories of diagrams $\mathit{Set}^{\mathbf{C}}$. They are categories defined in terms of *sites*, which are pairs consisting of a small category $\mathbf{C}$ and a Grothendieck (pre-)topology J on it. This means the following for a small category $\mathbf{C}$ with finite limits.

A *Grothendieck (pre-)topology* J on $\mathbf{C}$ is a family $J(A)$ of morphisms with A as codomain, called 'coverings of A', for each object A of $\mathbf{C}$, satisfying the following:

- Each singleton $\{\mathrm{id}_A\colon A \longrightarrow A\}$ belongs to $J(A)$.
- Coverings are stable under change of base (pullbacks).
- J is closed under composition. Given $\{f_i\colon A_i \longrightarrow A\}_{i\in I} \in J(A)$ and, for each $i \in I$, $\{g_{ik}\colon A_{ik} \longrightarrow A_i\}_{k\in I_i} \in J(A_i)$, then

$$\{f_i \circ g_{ik}\colon A_{ik} \longrightarrow A\}_{k\in I_i,\, i\in I} \in J(A).$$

Given a site $(\mathbf{C}, J)$, a *sheaf* on it is any functor $F\colon \mathbf{C}^{\mathrm{op}} \longrightarrow Set$ such that for every covering family $\{f_i\colon A_i \longrightarrow A\} \in J(A)$ and a compatible family of elements $a_i \in F(A_i)$, there exists a unique $a \in F(A)$ such that, for each $i \in I$, its restriction to A_i (image under $F(f_i)$) is a_i. Denote by $\mathrm{Sh}(\mathbf{C}, J)$ the full subcategory of the category $Set^{\mathbf{C}^{\mathrm{op}}}$ whose objects are sheaves on the site $(\mathbf{C}, J)$. That means that its morphisms are natural transformations between sheaves regarded as presheaves. By a *Grothendieck topos* it is meant any category of the form $\mathrm{Sh}(\mathbf{C}, J)$.

Along with the notion of a topos it is necessary to determine what notion of morphism between toposes to consider. Actually there are two, depending on what aspects of a topos one is focusing on.

The logical properties of a topos lead to the notion of a logical morphism.

Definition 1.10 Given toposes $\mathscr{E}$ and $\mathscr{F}$, a *logical morphism*

$$\mathscr{E} \longrightarrow \mathscr{F}$$

is any functor which preserves the topos structure, that is, finite limits and power objects, and therefore also Ω, exponentiation, finite colimits, and in fact, interpretations of first-order formulas whose sorts of the free variables are interpreted as objects of the topos.

If $\mathscr{E}$ is a topos, given any object X in $\mathscr{E}$, we can consider the slice category $\mathscr{E}/X$ whose objects are morphisms in $\mathscr{E}$ with codomain X and whose morphisms from $a\colon A \to X$ to $b\colon B \to X$ are morphisms $g\colon A \to B$ of $\mathscr{E}$ over X, in the sense that the triangle

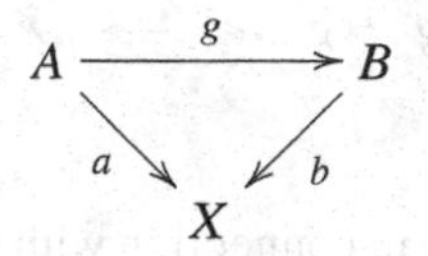

commutes. The category $\mathscr{E}/X$ is a topos. As a simple example of a logical morphism one has

$$X^*\colon \mathscr{E} \longrightarrow \mathscr{E}/X$$

for any object X of $\mathscr{E}$, which is given by pulling back along the unique morphism $X \to 1$ from X into the terminal object.

More generally, if $f\colon X \to Y$ in $\mathscr{E}$, then there is induced a functor

$$f^*\colon \mathscr{E}/Y \longrightarrow \mathscr{E}/X$$

defined by taking pullbacks along f. It is a logical functor. Without any further assumptions, there is a left adjoint

$$\Sigma_f \dashv f^*,$$

where Σ_f is defined by composition with f. In a topos, colimits are universal, that is, preserved by pullback functors. Moreover, the additional properties of a topos (exponentiation) imply that there is a right adjoint

$$f^* \dashv \Pi_f .$$

Among other examples of constructions involving logical morphisms are glueing, the ultrapower construction, and the free topos [55].

Thinking of toposes as generalized spaces (even though they may have no points) a notion of geometric morphism emerges by abstracting properties of continuous mappings. A continuous mapping

$$g \colon X \longrightarrow Y$$

lifts to the generalized spaces $\mathrm{Sh}(X)$ and $\mathrm{Sh}(Y)$ in the form of a functor

$$g^* \colon \mathrm{Sh}(Y) \longrightarrow \mathrm{Sh}(X)$$

given by pulling back the local homeomorphism along g. The functor g^* preserves finite limits and has a right adjoint g_*. This point of view motivates the following definition.

Definition 1.11 A *geometric morphism* from a topos $\mathscr{E}$ to a topos $\mathscr{F}$ is given by a pair (g^*, g_*), where

$$g^* \dashv g_* \colon \mathscr{E} \longrightarrow \mathscr{F}$$

and g^* preserves finite limits.

Geometric morphisms arise in connection with topologies in any topos $\mathscr{E}$, an abstraction of Grothendieck topologies due to Lawvere and Tierney. We recall the definition.

Definition 1.12 ([55]) Let $\mathscr{E}$ be a topos. A *topology* in $\mathscr{E}$ is a morphism $j \colon \Omega \longrightarrow \Omega$ such that the following diagrams are commutative :

(i)

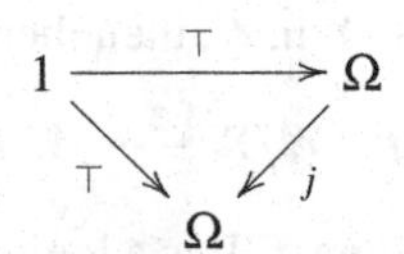

(ii)

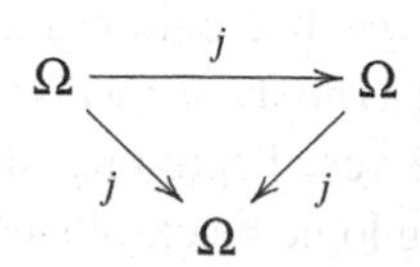

(iii)

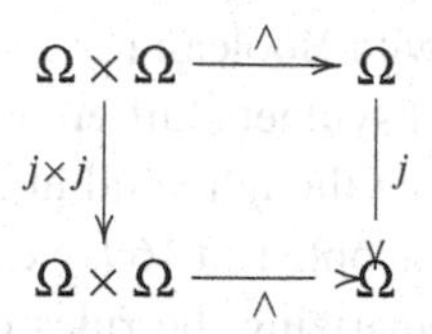

A topology j in a topos $\mathscr{E}$ gives rise to a new topos $\mathrm{Sh}_j(E)$ defined over $\mathscr{E}$ [55]. They were originally conceived by F. W. Lawvere as modal operators of the sort 'it is locally the case that ...'.

Remark 1.13 The expression 'elementary topos' has been used instead of simply 'topos' in order to emphasize the fact that toposes are described in an elementary or first-order language However, this use is a grammatical mistake. The correct term is 'topos' as already presented in [73] and [56]. The geometrical need existed for a general 2-category such that (mathematics could be carried out internally in it and) the categories U arising from Grothendieck universes would be objects, and applying Grothendieck's relativization $(-)/\mathscr{S}$ would capture Grothendieck's notion of U-topos when $\mathscr{S}$ is taken to be U. Thus, 'elementary topos' is not a technical term but rather an explanatory phrase used in certain contexts to emphasize that one is not presuming a strong set theory.

Remark 1.14 That the internal logic of the toposes of synthetic differential geometry must be Heyting (or intuitionistic) is well explained by Lavendhomme [69] by analysing the 'catastrophic' effect that adopting Boolean (or classical) logic would have on the basic axiom of synthetic differential geometry, to wit, the Kock-Lawvere axiom. This axiom may naively be stated to say that for every $f\colon D \longrightarrow R$ there exists a unique $b \in R$ such that for every $d \in D$, $f(d) = f(0) + d \cdot b$. If Boolean logic were employed, such a ring R would be the null ring. We first show that $D = \{0\}$. Indeed, let $f\colon D \longrightarrow R$ be described by $f(0) = 0$ and $f(d) = 1$ for every non-zero $d \in D$ (Boolean logic is already employed to define such an f). For this f there exists a unique $b \in R$ such that for every $d \in D$, $f(d) = d \cdot b$. If there were a non-zero $d \in D$ then we would have $1 = d \cdot b$ and so, multiplying by d, we would in turn get $d = 0$ (since $d^2 = 0$). This would lead to a contradiction, hence (once again using

Boolean logic) $D = \{0\}$. Now, apply this to the zero function $f : D \longrightarrow R$, where $D = \{0\}$. From $0 = 0 \cdot b$ with a unique such $b \in R$ follows that $b = 0$ for every $b \in R$. That is, $R = \{0\}$. The theory would then stop right after it began. We are then forced, from the very beginning of the theory, to make sure that only the principles of Heyting logic be employed.

On account of the considerations made above, one could perhaps simply warn the reader acquainted with Boolean (or classical) logic about the unsuitability for the development of synthetic differential geometry of certain principles and rules of inference that though valid in Boolean are not so in Heyting logic. This is the approach adopted in [69, Sect. 1.1.2], leaving for the end of the book the task of summarizing the rules of Heyting logic while urging 'the courageous reader' to check along the way that only these rules have been employed.

We could in a similar vein simply state that the logic that we need, weaker than Boolean (or classical) logic and allowing less deductions, is essentially a logic which differs from the classical in that it does not admit the law of excluded middle. In other words, we could just declare that the statement

$$\varphi \vee \neg \varphi$$

(where φ is any formula of the language) is not allowed for the practical reason that we wish to have models. In particular then, we could point out that the double negation principle

$$\neg(\neg \varphi) \Rightarrow \varphi$$

is not allowed in general either although in certain cases it is.

For the record, we list next some *not universally valid* deductions in the deductive system corresponding to Heyting predicate calculus. This does not mean that they should never be used, only that one must make sure that they are valid in the particular cases where they are used.

$$\frac{\neg\neg\varphi}{\varphi} \qquad \frac{\neg(\varphi \wedge \psi)}{\neg\varphi \vee \neg\psi} \qquad \frac{\forall a \in A\ (\varphi \vee \psi(a))}{\varphi \vee (\forall a \in A\ \psi(a))}.$$

In spite of the above considerations, we shall, in addition, give a list of *valid formulas* and *valid rules of inference* in Heyting logic that can be checked (if in doubt) along the way.

A list of *valid formulas* in Heyting logic follows, where the letters φ, ψ and χ denote formulas of the language.

$$\varphi \Rightarrow (\psi \Rightarrow \varphi) \qquad (\varphi \Rightarrow \psi) \Rightarrow \left[(\varphi \Rightarrow (\psi \Rightarrow \chi)) \Rightarrow (\varphi \Rightarrow \chi)\right]$$
$$\varphi \wedge \psi \Rightarrow \varphi \qquad \varphi \wedge \psi \Rightarrow \psi$$
$$\varphi \Rightarrow \varphi \vee \psi \qquad \psi \Rightarrow \varphi \vee \psi$$
$$\varphi \Rightarrow (\psi \Rightarrow \varphi \wedge \psi) \qquad (\varphi \Rightarrow \chi) \Rightarrow \left[(\psi \Rightarrow \chi) \Rightarrow (\varphi \vee \psi \Rightarrow \chi)\right]$$
$$\neg\varphi \Rightarrow (\varphi \Rightarrow \psi) \qquad (\varphi \Rightarrow \psi) \Rightarrow \left[(\varphi \Rightarrow \neg\psi) \Rightarrow \neg\varphi\right].$$

Also valid are the formulas

$$\varphi(t) \Rightarrow \exists a \in A\ \varphi(a) \qquad \text{and} \qquad \forall a \in A\ \varphi(a) \Rightarrow \varphi(t),$$

where t denotes a term in the language with no free variable bounded by quantifiers which bind the variable a in φ.

The *rules of inference* valid in Heyting logic can be taken to be the following:

$$\frac{\varphi \wedge (\varphi \Rightarrow \psi)}{\psi} \qquad \frac{\chi \Rightarrow \varphi(a)}{\chi \Rightarrow \forall a \in A\ \varphi(a)} \qquad \frac{\varphi(a) \Rightarrow \chi}{\exists a \in A\ \varphi(a) \Rightarrow \chi}$$

where χ is any formula in which a is not a free variable.

In addition, the following formulas involving equality are assumed to be valid in any topos $\mathscr{E}$:

$$a = a \qquad \text{and} \qquad (a = a') \Rightarrow (\varphi(a) \Rightarrow \varphi(a')).$$

Finally, the formula

$$(\varphi(0) \wedge \forall n \in N\, (\varphi(n) \Rightarrow \varphi(n+1))) \Rightarrow \forall n \in N\, \varphi(n)$$

expresses the principle of induction for any formula φ in which n is a variable of sort N, where $\langle N, s, 0 \rangle$ is a natural numbers object in $\mathscr{E}$.

Associated with the Heyting predicate calculus is a deductive system. Let Γ be a finite set of formulas and φ a single formula in the language. Denote by

$$\Gamma \vdash \varphi$$

the statement that the formula φ is deducible from the set of formulas in Γ. Rules of inference can be set up from the valid formulas already listed by means of the following metatheorem.

Theorem 1.15 *[58]. Let Γ be a finite set of formulas and φ, ψ any single formulas in the language. Then*

$$(\Gamma \wedge \varphi) \vdash \psi \qquad \textit{implies} \qquad \Gamma \vdash (\varphi \Rightarrow \psi).$$

In the deductive system corresponding to the Heyting predicate calculus, the following principles are instances of valid deductions

$$\begin{array}{c} \varphi \\ \hline \neg\neg\varphi \end{array} \qquad \begin{array}{c} \varphi \Rightarrow \neg\psi \\ \hline\hline \neg(\varphi \wedge \psi) \end{array} \qquad \begin{array}{c} \varphi \Rightarrow \psi \\ \hline \neg(\varphi \wedge \neg\psi) \end{array}$$

$$\begin{array}{c} \varphi \Rightarrow \psi \\ \hline \neg\psi \Rightarrow \neg\varphi \end{array} \qquad \begin{array}{c} \neg\varphi \vee \neg\psi \\ \hline \neg(\varphi \wedge \psi) \end{array} \qquad \begin{array}{c} \forall a \in A\ \neg\varphi(a) \\ \hline\hline \neg\exists a\ \varphi(a) \end{array}$$

$$\begin{array}{c} \exists a \in A\ \neg\varphi(a) \\ \hline \neg\forall a \in A\ \varphi(a) \end{array} \qquad \begin{array}{c} \exists a \in A\ \varphi(a) \\ \hline \neg\forall a \in A\ \neg\varphi(a) \end{array} \qquad \begin{array}{c} (\varphi \vee \psi) \Rightarrow \chi \\ \hline\hline (\varphi \Rightarrow \chi) \wedge (\psi \Rightarrow \chi) \end{array}$$

where the double line indicates validity in both directions and the single line means validity from top to bottom.

Exercise 1.16 Using the list of valid deductions given above, prove that

$$\begin{array}{c} (\bigvee_{i=1}^{n} \varphi_i) \Rightarrow \neg(\bigwedge_{i=1}^{n} \psi_i) \\ \hline \bigwedge_{i=1}^{n} (\varphi_i \Rightarrow \neg\psi_i) \end{array}$$

is also a valid deduction in Heyting logic.

Remark 1.17 If $\varphi \vdash \psi$ is valid in Heyting logic, then it is also classically valid. However, from $\varphi \vdash \psi$ classically valid, one can only deduce that $\varphi \vdash \neg\neg\psi$ is valid in Heyting logic. Notice, however that, using the first and fourth of the valid principles listed above, in order to prove a formula of the form $\varphi \vdash \neg\psi$, it is enough to prove $\psi \vdash \neg\varphi$. In particular, if a formula $\varphi \vdash \neg\psi$ is classically valid, then it is also valid in Heyting logic.

Exercise 1.18 Prove, using Heyting logic, that for any topos $\mathscr{E}$,

$$\neg\neg : \Omega \longrightarrow \Omega$$

is a topology in the sense of Definition 1.12.

The topos corresponding to it, $\mathrm{Sh}_{\neg\neg}(X)$ of 'double negation sheaves', is a Boolean topos (i.e., a topos in which the Heyting algebra Ω is a Boolean algebra, hence logic becomes classical therein) [55].

Coherent logic is a fragment of finitary first-order logic (with equality) which allows only the connectives $\wedge$, $\vee$, $\top$, $\perp$, $\exists$. It is usually presented in terms of *sequents*

$$\varphi \vdash \psi$$

with φ and ψ coherent formulas in possibly n free variables $x_1, \ldots, x_n$. In full first-order logic, such a sequent is equivalent to a single formula, to wit

$$\forall x_1 \cdots \forall x_n (\varphi \Rightarrow \psi).$$

This means that, in these terms, one occurrence of $\Rightarrow$ is allowed inside the formula and any finite string of universal quantifiers can occur at the outer level of the formula.

For instance, for a commutative ring A with unit, where $a \in A^* \subset A$ is the extent of the formula $\exists x \in A\ (a \cdot x = 1)$, the formula

F1. $\forall a \in A\, (a = 0 \vee a \in A^*)$

is coherent and expresses the property of A being a field. On the other hand, the classically (but not intuitionistically) equivalent formulas

F2. $\forall a \in A\, (\neg(a = 0) \Rightarrow a \in A^*)$

and

F3. $\forall a \in A\, (\neg(a \in A^*) \Rightarrow a = 0)$

are not coherent.

Remark 1.19 With the addition of the assumption $\neg(0 = 1)$ for a commutative ring A with unit 1 in a topos $\mathscr{E}$, F1 is often referred to as a 'geometric field' since it is expressed in the geometric language of the theory of rings, in the sense of formulas preserved by the inverse image parts of geometric morphisms. Geometric logic differs from coherent logic only in that infinitary disjunctions are allowed in addition to finite ones. Alternatively, coherent logic may be viewed as finitary geometric logic.

Also geometric (in fact, coherent) is the notion of an *integral domain* which, for a commutative ring A with unit is expressed by the formula

$$\forall a \in A\ \left[a \cdot a' = 0 \Rightarrow (a = 0) \vee (a' = 0)\right].$$

If A is an integral domain, then the subobject $U \rightarrowtail A$ of its non-zero elements is multiplicatively closed, and the ring $A\left[U^{-1}\right]$ is a *field of fractions* in the sense that it satisfies F2 [55], although the terminology refers to the construction that renders all elements of $U \rightarrowtail A$ universally invertible in it.

Dually, if A is a *local ring* meaning that A is a commutative ring with unit, which in addition satisfies the formula

$$\forall a \in A \; [a \in A^* \vee (1-a) \in A^*]\,,$$

then the subobject $M \rightarrowtail A$ of its non-invertible elements is an ideal, and the quotient A/M is a *residue field* in the sense that it satisfies F3.

What is important about a coherent theory, that is a theory whose axioms are all given by coherent formulas, is that models in *Set* are enough to test their validity in any topos. We refer the reader to [90] for a full account of coherent theories and their 'classifying toposes'. The completeness theorem just informally stated for coherent theories is a consequence of theorems of Deligne and Barr to the effect that the classifying toposes of coherent theories ('coherent toposes') have enough points [55, 6].

1. Given a topos $\mathscr{E}$ with a natural numbers object N, the *object of integers* Z can be constructed in $\mathscr{E}$ by means of the pushout

$$\begin{array}{ccc} 1 & \xrightarrow{0} & N \\ {\scriptstyle 0}\downarrow & & \downarrow{\scriptstyle \rho_2} \\ N & \xrightarrow[\rho_1]{} & Z \end{array}$$

The *object of the rationals* Q can be constructed as the *ring of fractions* $Z[P^{-1}]$, where

$$P = [[z \in Z \mid \exists n \in N\,(z = \rho_1(s(n))]]$$

is the subobject of positive integers, which is a multiplicative set.

The objects Z and Q are preserved by inverse image parts of geometric morphisms. A different matter is that of an object of real numbers in $\mathscr{E}$, since classically equivalent constructions may give non-isomorphic objects. For instance, there is an object R_d (*Dedekind reals*) and an object R_c (*Cauchy reals*) both candidates for an object of real numbers in $\mathscr{E}$, but these two are in general not isomorphic objects.

2. A consequence of using Heyting (rather than Boolean) logic is that it forces mathematicians to be more discriminating in their choice of concepts. For instance, to say that a non-trivial commutative ring A is a field, we may, using Boolean logic, choose from any of the above listed equivalent notions. In fact, there are interesting 'field-like' objects in any topos with a natural numbers object which, such as the Dedekind reals R_d, satisfy F2 or F3 but

not F1. In effect, F2 corresponds to fields which arise as fields of fractions of integral domains, whereas F3 corresponds to those fields that arise as residue fields of local rings. For a prime ideal $\mathfrak{p}$, the field of fractions of an integral domain $A/\mathfrak{p}$ equals the residue field of the local ring $A_{\mathfrak{p}}$, and either one yields a field satisfying F1.

3. Another construction, this time involving the *axiom of choice*, is that of the *algebraic closure* of a field. Even for geometric fields, algebraic closures need not exist in a topos, although, for a Grothendieck topos $\mathscr{E}$, a theorem of Barr [55, 5] says that there always exist a topos $\mathscr{F}$ satisfying the axiom of choice and a surjection $f\colon \mathscr{F} \longrightarrow \mathscr{E}$ (i.e., a geometric morphism f for which f^* is faithful) so that, by 'moving' to $\mathscr{F}$ we can find all algebraic closures—the problem being that in general, and in view of the poor preservation properties of the inverse image functor f_*, it is not possible to come back to $\mathscr{E}$ without losing much of what was gained. By contrast, the theory of ordered fields has a more constructive flavour. One aspect of this is the fact, proven in [19], that in any Grothendieck topos $\mathscr{E}$, the *real closure* of an ordered field always exists.

4. Although we have concentrated on algebra in the above examples, other branches of mathematics, such as general topology and analysis, have been for a long time the object of interest for constructivist mathematics [50, 41]. For example, André Joyal [57] has constructed a topos in which the unit interval (of the Dedekind reals) is not compact. Hence, such a property of the real numbers cannot be established within set theory without the axiom of choice. An interesting (and accessible) article of the virtues of constructive mathematics is that of Andrej Bauer [7]. In this connection we also refer to [21] for selected applications of topos theory in logic, set theory, and model theory.

1.2 Logico-Geometric Notions in Toposes

In the study of synthetic differential geometry and topology two additional notions arose in the context of toposes—to wit, atoms and Penon objects. We discuss them briefly in this section. More details will be given in the rest of the book.

If a topos $\mathscr{E}$ is constructed from topology it will be often the case that one or more topological structures can be put on its objects. What is remarkable is that, even if nothing special is assumed about a topos $\mathscr{E}$, its intrinsic logic is enough to produce a topological structure[1] on its objects. This interesting idea is due to Jacques Penon [96].

[1] A proper definition of 'topological structure' in a topos is postponed until Chapter 5.

Definition 1.20 Let $\mathscr{E}$ be any topos and X an object of $\mathscr{E}$. For $U \subset X$ in $\mathscr{E}$, U is said to be an *intrinsic (or Penon) open* of X if $\mathscr{E}$ satisfies

$$\forall x \in X\, \forall y \in X\, [x \in U \Rightarrow (\neg(y = x) \vee y \in U)].$$

Exercise 1.21 Let $U \subset X$ be an intrinsic (or Penon) open in the sense of Definition 1.20. Prove that the following holds.

$$\forall x \in U\, (\neg\neg\{x\} \subset U).$$

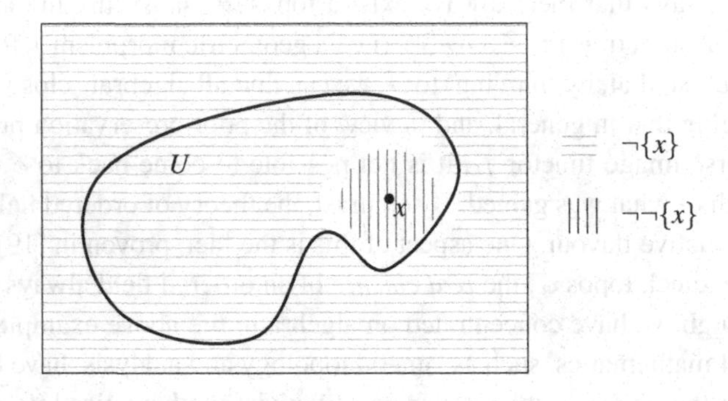

Figure 1.1 Penon open object

Proposition 1.22 *Let $\mathscr{E}$ be any topos.*

(i) *For any object X of $\mathscr{E}$, the subobject $P(X) \subset \Omega^X$ of all intrinsic (or Penon) opens of X is closed under arbitrary suprema and finite infima in Ω^X.*

(ii) *The property of being an intrinsic (or Penon) open is (i) stable under inverse images of morphisms in $\mathscr{E}$, that is, for any $f\colon Y \to X$ and $U \in P(X)$ it follows that $f^{-1}(U) \in P(Y)$, and (ii) closed under composition, that is, if $U \subset X$ is a Penon open of X, and $V \subset U$ is a Penon open of U, then $V \subset X$ is a Penon open of X.*

Exercise 1.23 Using Heyting logic alone prove both statements of Proposition 1.22.

In synthetic differential geometry and topology, certain objects are implicitly thought of as 'infinitesimals' by virtue of the axioms of jets and of germs representability. It seemed desirable, however, to pin down this idea by means of an additional axiom in each case. In this section we lay down the definitions that will be employed later in the axiomatic treatments of differential geometry and topology and explore some of their aspects. Our sources for this section are [16], [76], [42], and [113].

Definition 1.24 An object A of $\mathscr{E}$ is said to be

(i) an *atom* if the endofunctor $(-)^A : \mathscr{E} \longrightarrow \mathscr{E}$ has a right adjoint, denoted $(-)_A$,
(ii) *tiny* if A is well supported (in the sense that $A \twoheadrightarrow 1$ in $\mathscr{E}$) and an atom,
(iii) *infinitesimal* if A is an atom and has a unique point (global section).

Remark 1.25 The notion of an atom, as in Definition 1.24, is due to F. W. Lawvere. It arose in connection with a characterization of categories of diagrams [16] given by analogy with the characterization of fields of sets as complete atomic Boolean algebras [105]. Therein, by an atom (or a *Set*-atom) of a regular category $\mathscr{E}$, it was meant an object A of $\mathscr{E}$ for which the functor $\mathrm{Hom}_{\mathscr{E}}(A,-) : \mathscr{E} \twoheadrightarrow Set$ preserves colimits. In effect, by regularity in the sense of [16], this means that A is both projective and (infinitary) connected. Related versions of the notion of an atom were given in [17], [18], and [76]. The terminology 'tiny' was suggested by Freyd [42] to replace that of 'atom' used in those sources in order to avoid conflict with another notion by that name, due to Barr [5]. However, in view of the main result in [113], the intended meaning should have been that of a well-supported atom, which is precisely what we here have meant by 'tiny'. For historical reasons we keep both here. The instances of atoms that arise in synthetic differential geometry are tiny by virtue of their being in fact infinitesimal.

In $\mathscr{E} = Set$, only the terminal object 1 is an atom and it is tiny. In a *diagrammatic category* $\mathscr{E} = Set^{\mathbf{C}}$, of *Set*-valued functors on a small category $\mathbf{C}$, if $\mathbf{C}$ has finite coproducts, then every representable object is an atom. Explicitly, for any $F \in Set^{\mathbf{C}}$, the exponential

$$\begin{aligned} F^{\overline{B}}(B') &= \mathrm{Hom}_{\mathscr{E}}(\overline{B'} \times \overline{B}, F) \\ &\simeq \mathrm{Hom}_{\mathscr{E}}(\overline{B' + B}, F) \simeq F(B' + B) \\ &\simeq [F \circ (B + (-))](B'), \end{aligned}$$

where $\overline{B}$ stands for $\mathrm{Hom}_{\mathbf{C}}(B,-)$, for B an object in $\mathbf{C}$. Since $(-)^{\overline{B}}$ is induced by composition with $B + (-)$, it has a right adjoint, denoted $(-)_{\overline{B}}$.

Categories of diagrams of the form $Set^{\mathbf{C}^{\mathrm{op}}}$ are referred to as *presheaf toposes*. Not every atom in an arbitrary presheaf topos is tiny. Consider a non-trivial topological space X and let $\mathscr{O}(X)$ be the partially ordered set of open subsets of X with inclusions as morphisms. The representable functors in $Set^{\mathscr{O}(X)^{\mathrm{op}}}$ are the subobjects of the terminal object, hence the only well supported such is the terminal object itself, that is $1 = \widetilde{X}$.

For a *Set*-based *Set*-cocomplete category $\mathscr{E}$, an object A of $\mathscr{E}$ is said to be a *Set*-atom [16] (see [17] for the version relative to a closed category $\mathscr{V}$) if the functor $\mathrm{Hom}_{\mathscr{E}}(A,-)\colon \mathscr{E} \longrightarrow Set$ preserves all *Set*-colimits. The connection between the intrinsic notion of an atom given in Definition 1.24 and that of a *Set*-atom is the following. For a *local* Grothendieck topos $\mathscr{E}$, the *Set*-atoms agree with those of Definition 1.24. The sufficiency was observed in [23] : if $\mathscr{E}$ is a local Grothendieck topos and A is an atom in $\mathscr{E}$ in the sense of Definition 1.24, then the composite

$$\mathscr{E} \xrightarrow{(-)^A} \mathscr{E} \xrightarrow{\Gamma} Set$$

has a right adjoint since each functor in the composite does, hence preserves all *Set*-colimits. For the converse stating that for a local Grothendieck topos $\mathscr{E}$, if an object A of $\mathscr{E}$ is a *Set*-atom then it is also an atom in the sense of Definition 1.24, use the adjoint triangle theorem of [31]. In synthetic differential geometry it is often the case that atoms are used in an external rather than internal sense [64]. Uses of the intrinsic notion of atom, in particular useful interpretations of the 'amazing right adjoint' $(-)_A$, are still being explored [66].

Definition 1.26 Let $\mathscr{E}$ be a topos and let A be a tiny object of $\mathscr{E}$. The object X is said to be *A-discrete* if the monomorphism $X^p\colon X \rightarrowtail X^A$ is an epimorphism (hence an isomorphism). Denote by $\mathscr{T}_A$ the full subcategory of $\mathscr{E}$ whose objects are the A-discrete objects of $\mathscr{E}$.

Remark 1.27 The notation $\mathscr{T}_A$ for A a tiny object of the given topos $\mathscr{E}$ indicates that $\mathscr{T}_A$ will play the role of a category of 'test objects' for $\mathscr{E}$. For this reason Lawvere [75] proposed this notation rather than the alternative $\mathscr{D}_A$ for 'A-discrete'.

The following theorem—conjectured by Lawvere [72] and proved by Freyd [42] (see also [113])—exhibits an important role of the tiny objects in a topos.

Theorem 1.28 *If A is a tiny object in a topos $\mathscr{E}$, then the full subcategory $\mathscr{T}_A$ is both reflective and coreflective in $\mathscr{E}$, hence it is a topos and there is an essential geometric morphism*

$$\gamma\colon \mathscr{E} \longrightarrow \mathscr{T}_A$$

whose inverse image part is the inclusion. Moreover, $\mathscr{T}_A$ is an exponential ideal and in particular the reflection preserves finite products. If $\mathscr{E}$ has a natural numbers object N then so does $\mathscr{T}_A$ since N is A-discrete.

Remark 1.29 The terminal object 1 in (any topos) $\mathscr{E}$ is tiny. In this case, every object X of $\mathscr{E}$ is 1-discrete so we gain nothing by considering $\mathscr{T}_1$. In the cases where $p\colon A \twoheadrightarrow 1$ is not an isomorphism in $\mathscr{E}$, the resulting full subcategory $\mathscr{T}_A$ of $\mathscr{E}$ is non-trivial. Consider $\Omega \in \mathscr{E}$. We claim that $\Omega \notin \mathscr{T}_A$. Indeed, it is well known that $(-)^{\Omega}\colon \mathscr{E} \longrightarrow \mathscr{E}$ is monadic (Paré's theorem [92]) and so in particular it reflects isomorphisms. Hence the assumption $\Omega^p\colon \Omega \rightarrowtail \Omega^A$ an isomorphism would imply that $p\colon A \twoheadrightarrow 1$ iso, a contradiction. The basic axioms of synthetic differential geometry and topology force the tiny objects representing jets and germs to be non-trivial, that is, not isomorphic to the terminal object. For the same reason and assuming A non-trivially tiny, A is not A-discrete. To see this, denote by

$$\check{(-)}\colon \mathscr{E} \longrightarrow \mathscr{T}_A$$

the right adjoint to the inclusion

$$\iota\colon \mathscr{T}_A \hookrightarrow \mathscr{E}$$

which exists (and is explicitly constructed in the proof) by Theorem 1.28. If A were A-discrete, $\check{\Omega}^A$ would also be A-discrete since $\mathscr{T}_A$ is a topos, hence $\check{\Omega} \cong \check{\Omega}^A$ would imply that $\check{\Omega}$ is A-discrete. Therefore, from $\check{\Omega}^p\colon \check{\Omega} \rightarrowtail \check{\Omega}^A$ iso would follow that $p\colon A \twoheadrightarrow 1$ is an isomorphism, a contradiction.

Remark 1.30 The notion of an atom (as stated in Definition 1.24) is problematic within topos theory for the following reason. Although for any object A of a topos $\mathscr{E}$, the adjunction

$$(-)\times A \dashv (-)^A$$

is *strong* in the sense that for every objects X, Y, B of $\mathscr{E}$ there is a bijection

$$\frac{B \longrightarrow Y^{(X\times A)}}{B \longrightarrow (Y^A)^X}$$

since $\mathscr{E}$ is cartesian closed, it is not true in general that the objects $X^{(Y^A)}$ and $(X_A)^Y$, for A an atom, are isomorphic. Investigating this more closely one finds that by restricting to the A-discrete objects T of $\mathscr{E}$, the adjointness $(-)^A \dashv (-)_A$ is strong. Indeed, one has the following bijections

$$\frac{\frac{\frac{\frac{\frac{T \longrightarrow X^{(Y^A)}}{T \times Y^A \longrightarrow X}}{T^A \times Y^A \longrightarrow X}}{(T \times Y)^A \longrightarrow X}}{(T \times Y) \longrightarrow X_A}}{T \longrightarrow (X_A)^Y}$$

where the A-discreteness of T is used to pass from the second to the third row.

The following consequence of Theorem 1.28, due to Yetter [113], gives a general reason why tiny objects are preserved by the associated sheaf functor $\mathrm{Sh}_j(\mathscr{E}) \hookrightarrow \mathscr{E}$, whereas atoms need not be so preserved, not even when the topology $j\colon \Omega \longrightarrow \Omega$ in $\mathscr{E}$ is subcanonical (representable functors are sheaves).

Lemma 1.31 *Let $\mathscr{E}$ be a topos, A a tiny object in $\mathscr{E}$, and j a topology in $\mathscr{E}$, with corresponding topos $\mathrm{Sh}_j(\mathscr{E})$ of j-sheaves and inclusion*

$$\iota\colon \mathrm{Sh}_j(\mathscr{E}) \hookrightarrow \mathscr{E}.$$

Then the following square is a pullback

$$\begin{array}{ccc} \mathrm{Sh}_j(\mathscr{E}) & \longrightarrow & \mathrm{Sh}_{\check{j}}(\mathscr{T}_A) \\ \downarrow & \lrcorner & \downarrow \\ \mathscr{E} & \longrightarrow & \mathscr{T}_A \end{array}$$

By functoriality of the right adjoint to the inclusion $\iota\colon \mathrm{Sh}_j(\mathscr{E}) \hookrightarrow \mathscr{E}$ and the fact that it preserves products and the terminal objects, it follows that since $j\colon \Omega \to \Omega$ is a topology, so is $\check{j}\colon \check{\Omega} \to \check{\Omega}$. It follows from Lemma 1.31 that the essential geometric endomorphism φ_A of $\mathscr{E}$ given by

$$(-) \times A \dashv (-)^A \dashv (-)_A$$

lifts uniquely to an essential geometric endomorphism ψ_A of $\mathrm{Sh}_j(\mathscr{E})$ as in the

prism

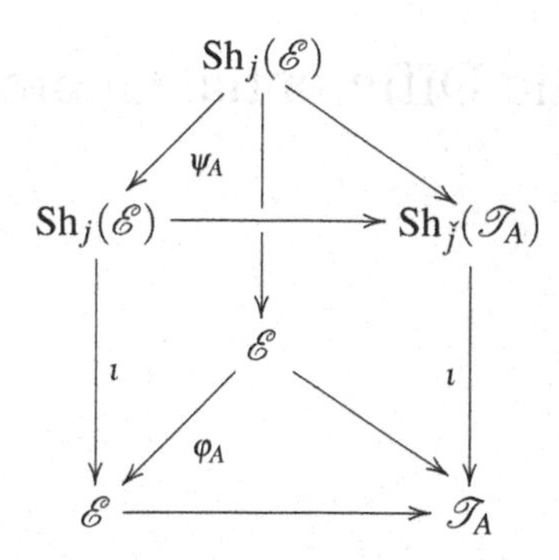

It is easily seen that $\psi_A{}^* = (-)^{\iota^* A}$.

Theorem 1.32 *Let $\mathscr{E}$ be a topos, A a tiny object in $\mathscr{E}$ and j a topology in $\mathscr{E}$, with corresponding topos* $\mathrm{Sh}_j(\mathscr{E})$ *of j-sheaves and inclusion* $\iota : \mathrm{Sh}_j(\mathscr{E}) \hookrightarrow \mathscr{E}$. *Then $\iota^* A$ is a tiny object of* $\mathrm{Sh}_j(\mathscr{E})$.

Proof It follows from Lemma 1.31 and the remarks made after it that $\iota^* A$ is an atom in $\mathrm{Sh}_j(\mathscr{E})$. It is also well supported since ι^*, which has a right adjoint and preserves finite limits, preserves both epimorphisms and the terminal object. Hence $\iota^* A$ is tiny in $\mathrm{Sh}_j(\mathscr{E})$. □

We end this chapter by posing some problems concerning the notions introduced in this section.

Exercise 1.33

(i) Prove that the result of Theorem 1.32 is still valid if 'tiny' is replaced by 'infinitesimal' in the sense of Definition 1.24.

(ii) As already remarked, there are toposes generated by atoms that are not (all of them) tiny. Investigate the nature of toposes generated by tiny objects that are not (all of them) infinitesimal.

(iii) Investigate what more one can say about the essential geometric morphism from Theorem 1.28 in the case where A is not just a tiny object of the topos $\mathscr{E}$ but an infinitesimal object of it.

2

Synthetic Differential Geometry

The data for synthetic differential geometry (SDG) is that of a pair $(\mathscr{E}, R)$ where $\mathscr{E}$ is a topos and R is a commutative ring with unit in it. The basic axioms of SDG are taken here to be the axiom of jets representability (Axiom J) and an axiom that states that the jets representing objects are tiny (well supported atoms) (Axiom W). By SDG we shall here mean any model $(\mathscr{E}, R)$ of these two basic axioms, which in addition satisfies that R is a field in the sense of Kock (Postulate K), that R satisfies the Reyes-Fermat condition (Postulate F) and that R is an Archimedean ordered ring (Postulate O).

2.1 The Axioms of SDG

All notions and axioms mentioned in this section can be seen in expanded form in [61]. For this reason, and since this first chapter is meant to be just introductory, we do not attempt to justify certain statements that can be found therein. Alternatively, such statements may be considered as exercises for the reader.

We begin by stating the axiom of representability of jet bundles, due, in its original form, to F. W. Lawvere and A. Kock. It was inspired by ideas of C. Ehresmann [40] and A. Weil [111].

Recall that a *Weil algebra* W is an algebra over the rational numbers $\mathbb{Q}$, equipped with a morphism

$$\pi : W \longrightarrow \mathbb{Q}$$

such that W is a local ring with maximal ideal

$$I = \pi^{-1}(0),$$

with I a nilpotent ideal, and such that W is a finite dimensional $\mathbb{Q}$-vector space.

The notion of Weil algebra makes sense in any topos $\mathscr{E}$ with a natural numbers object N. Recall from Chap. 1 that there is in $\mathscr{E}$ an object Q of rational numbers, constructed as the field of fractions $Q = Z[P^{-1}]$, where Z is the object of integers and P its subobject of positive integers. The constructions and properties of Weil algebras can be reproduced *verbatim* in any topos $\mathscr{E}$ with a natural numbers object. We use this remark in what follows.

For any Weil algebra W in a topos $\mathscr{E}$ with a natural numbers object N, the following statement holds internally in $\mathscr{E}$:

$$\exists ! \ell \in Z \left(W \simeq Q^{\ell+1} \right).$$

Let us fix a linear basis $\{e_0, e_1, \ldots, e_l\}$ for W as a Q-vector space. Consider next the matrix (γ_{ij}^k) with rational coefficients, obtained from the multiplication table presenting W. Given any Q-algebra A in $\mathscr{E}$, the matrix just depicted determines an A-algebra structure on A^{l+1}. Such an A-algebra will be denoted by $A \otimes W$ and it is isomorphic to A^{l+1}. It is the case that different presentations of W give the same A-algebra $A \otimes W$ and that the assignment $W \mapsto A \otimes W$ is functorial. Moreover, if $\{W \longrightarrow W_i\}$ is any finite inverse limit of Weil algebras, then the corresponding diagram $\{A \otimes W \longrightarrow A \otimes W_i\}$ is also an inverse limit.

Consider now given a commutative ring R with unit in a topos $\mathscr{E}$. Assume furthermore that R is a Q-algebra. Let W be a Weil algebra in $\mathscr{E}$ with presentation $\{h_i(X_1, \ldots, X_n)\}$. Denote by $\mathrm{Spec}_R(W)$ the zero set of the h_i in R^n, that is

$$\mathrm{Spec}_R(W) = [[(x_1, \ldots, x_n) \in R^n \mid \forall i \; h_i(x_1, \ldots, x_n) = 0]].$$

The restriction of a polynomial $\varphi \in R[X_1, \ldots, X_n]$ to $\mathrm{Spec}_R(W)$ defines a morphism

$$R[X_1, \ldots, X_n] \longrightarrow R^{\mathrm{Spec}_R(W)}$$

that sends each h_i to 0. Hence, it factors through the quotient of $R[X_1, \ldots, X_n]$ by the ideal generated by the h_i, that is, it factors through $R \otimes W$ via a unique morphism $\alpha \colon R \otimes W \longrightarrow R^{\mathrm{Spec}_R(W)}$.

Axiom 2.1 (Axiom J) (Jets Representability) *For each Weil algebra W, the ring morphism*

$$\alpha \colon R \otimes W \longrightarrow R^{\mathrm{Spec}_R(W)},$$

given by the rule

$$\psi \mapsto [p \mapsto \varphi(p)],$$

where $p \in \mathrm{Spec}_R(W)$ and φ is any polynomial that represents $\psi \in R \otimes W$, is invertible.

We shall next review and explain some particular cases of Axiom J which were historically considered before the more general form and led to it for categorical as well as practical reasons.

Let

$$D = [[x \in R \mid x^2 = 0]] .$$

Let W be the Weil algebra presented by $Q[\varepsilon] = Q[X]/(X^2)$. Then $R \otimes W$ admits a presentation of the form

$$R[\varepsilon] = R[X]/(X^2) ,$$

$R[\varepsilon]$ being the ring of dual numbers in $\mathscr{E}$. We have that $D = \mathrm{Spec}_R(W)$ and Axiom J in this case says precisely that the morphism of rings

$$\alpha \colon R[\varepsilon] \longrightarrow R^D$$

defined so that $\varphi(\varepsilon) \mapsto [d \mapsto \varphi(d)]$ is invertible. Equivalently, the axiom says that the morphism

$$\alpha \colon R \times R \longrightarrow R^D$$

given by the rule $(a,b) \mapsto [d \mapsto a + d \cdot b]$ is invertible. When convenient, we will refer to this particular case of Axiom J as 'Kock-Lawvere axiom'.

For any given $f \in R^R$ and $x \in R$, it follows that for all $d \in D$,

$$f(x+d) = f(x) + d \cdot b$$

for a unique b which depends on f as well as on x. This defines an element $f' \in R^R$, said to be the *derivative* of f, which is characterized by the equation

$$\forall d \in D \left[f(x+d) = f(x) + d \cdot f'(x) \right] .$$

The rules for derivatives follow from it by simple calculations. Thus we have:

$$\begin{aligned} &(f+g)' = f' + g', \\ &(a \cdot f)' = a \cdot f', \quad \text{for } a \in R \\ &(f \cdot g)' = f' \cdot g + f \cdot g', \quad \text{which is the Leibniz's rule.} \end{aligned}$$

We see then that the object D of $\mathscr{E}$ is so small that the graph of any function $f \colon R \longrightarrow R$ restricted to D is part of a straight line, but large enough so that this line is unique. That is, the restriction of $f \colon R \longrightarrow R$ to $D \subset R$ is completely characterized by the 1-jet (at 0) of f, that is, by the pair $(f(0), f'(0))$.

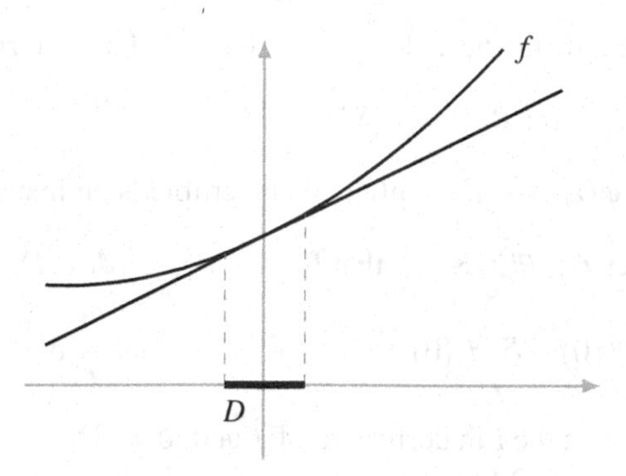

Figure 2.1 Kock-Lawvere axiom

Suppose now that we are given $f \in R^D$ instead of $f \in R^R$. By the above, $f'(0)$ is well defined, but the same cannot be said for elements $t \in D$ other than 0. This is because D is not closed under addition. In particular, we cannot, in this setting, define the second derivative of f at 0. However, for any $d_1, d_2 \in D$, $(d_1 + d_2)^3 = 0$ so that, if we know f on $D_2 = [[x \in R \mid x^3 = 0]]$, the second derivative of f at 0 can be defined as follows:

$$\begin{aligned} f(d_1 + d_2) &= f(d_1) + d_2 \cdot f'(d_1) \\ &= f(0) + d_1 \cdot f'(0) + d_2 \cdot (f'(0) + d_1 \cdot f''(0) \\ &= f(0) + (d_1 + d_2) \cdot f'(0) + (d_2 d_1) \cdot f''(0). \end{aligned}$$

Now, $(d_1 + d_2)^2 = 2d_1 d_2$, and therefore $d_1 + d_2 = \frac{1}{2} d_1 d_2$, provided that 2 is invertible, and we have

$$f(d_1 + d_2) = f(0) + (d_1 + d_2) \cdot f'(0) + \frac{1}{2}(d_2 + d_1) \cdot f''(0).$$

In the same vein, the derivative of any $f \in R^{D_r}$, where $D_r = [[x \in R \mid x^{r+1} = 0]]$ is defined on D_{r-1}. Let $U \subset R$ be such that $D_r \subset U$. It follows that for $f, g \in R^U$,

$$f|_{D_r} = g|_{D_r} \Rightarrow \left(f(0) = g(0) \wedge f'|_{D_{r-1}} = g'|_{D_{r-1}} \right).$$

Iterating the derivation we see that from $f|_{D_r} = g|_{D_r}$ follows that f and g have all same derivatives of order $i \leq r$ at 0.

Formally now, let

$$D_r = [[x \in R \mid x^{r+1} = 0]]$$

so that $D_r = \mathrm{Spec}_R(W_r)$, where $W_r = Q[X]/(X^{r+1})$. That the ring morphism

$$\alpha\colon R[\varepsilon_r] = R[X]/(X^{r+1}) \longrightarrow R^{D_r}$$

given by the rule $\varphi(\varepsilon_r) \mapsto [d \mapsto \varphi(d)]$, is invertible is an instance of Axiom J.

Exercise 2.2 Let $f \in R^{D_r}$. Show that for any $d_1, \ldots, d_r \in D$,

$$f(\delta) = f(0) + \delta \cdot f'(0) + \frac{1}{2}\delta^2 \cdot f''(0) + \cdots + \frac{1}{r!}\delta^r \cdot f^{(r)}(0),$$

where $f^{(i)}(0)$ denotes the i-th derivative of f at 0, $\delta = d_1 + \cdots + d_r$ and $1, \ldots, r$ are assumed to be invertible.

Denote by D_∞ the object of all nilpotent elements of R.

Lemma 2.3 *We have the following relations:*

(i) $D \subset D_2 \subset D_3 \subset \cdots \subset D_r \subset \cdots \subset D_\infty$

(ii) $D_\infty = \bigcup_{r \geq 1} D_r$

(iii) $D_\infty(2) = D_\infty \times D_\infty$

The ring of formal power series $R[[X]]$ and the object D_∞ of all nilpotent elements of R are related as a consequence of Axiom J, to wit: the ring morphism

$$\alpha\colon R[[X]] \longrightarrow R^{D_\infty},$$

given by the rule $\varphi(X) \mapsto [d \mapsto \varphi(d)]$, is invertible. Taylor's formula is easily established.

Proposition 2.4 *Given $U \subset R$ and $f \in R^U$,*

$$f(p+d) = P_p^\infty f(p+d) = \sum_{i=0}^{\infty} d^i \cdot \frac{1}{i!} f^{(i)}(p)$$

holds for all $d \in D_\infty$ and all $p \in U$ such that $p + D_\infty \subset U$.

Proof Since $d \in D_\infty$, there is an $r > 0$ such that $d \in D_r$. It follows now from Axiom J that there is a polynomial φ of degree at most r such that $f(p+x) = \varphi(x)$ for all $x \in D_r$. It follows in turn that $f^{(i)}(p) = \varphi^{(i)}(0)$ for all $i \leq r$, hence

$$P_p^\infty f(p+d) = P_0^\infty \varphi(d).$$

It is easy to see that for a polynomial of degree at most r, such as φ, $P_0^\infty \varphi = \varphi$. This finishes the proof. □

In order to continue developing the differential calculus, we need to be able to define partial derivatives. For any pair $n, r > 0$, consider

$$D_r(n) \subset R^n$$

consisting of those n-tuples $(x_1, \ldots, x_n)$ for which any product of $r+1$ elements taken from that list is zero. Clearly, $D_r(n) = \mathrm{Spec}(W_r^n)$, where W_r^n is the Weil algebra presented as $Q[X_1, \ldots, X_n]$ modulo the ideal generated by all products of $r+1$ of the symbols $X_1, \ldots, X_n$ with possible repetitions. Every element $\varphi \in R \otimes W_r^n$ is in a unique way a polynomial expression $\varphi(\varepsilon_1, \ldots, \varepsilon_n)$ where the product of any $r+1$ of the ε_i is zero. The restriction to $D_r(n)$ of a polynomial in n variables, considered as an element of R^{R^n}, defines a ring homomorphism

$$R[X_1, \ldots, X_n] \longrightarrow R^{D_r(n)}$$

such that the product of any $r+1$ of the X_i is sent to zero. This induces a unique ring homomorphism

$$\alpha \colon R \otimes W_r^n \longrightarrow R^{D_r(n)}$$

which is an isomorphism by virtue of Axiom J.

This instance of the axiom leads to partial derivatives. We illustrate it in the case $n = 2$, $r = 1$. Let $f \in R^{R^2}$. For a fixed $(x_1, x_2) \in R^2$, consider $g, h \in R^D$ defined and expanded by using Axiom J, as

$$g(d) = f(x_1 + d, x_2) = f(x_1, x_2) + d \cdot \frac{\partial f}{\partial x_1}(x_1, x_2)$$

and

$$h(d) = f(x_1, x_2 + d) = f(x_1, x_2) + d \cdot \frac{\partial f}{\partial x_2}(x_1, x_2).$$

If f is not defined on all of R^2 but on a subobject $U \subset R^2$, it is still possible to define the partial derivatives on the subobject of U given as follows:

$$[[(x_1, x_2) \in U \mid \forall d \in D \; ((x_1 + d, x_2) \in U \wedge (x_1, x_2 + d) \in U)]].$$

This explains why in order to have such partial derivatives it is enough to know f restricted to $D(2) = [[(x_1, x_2) \mid {x_1}^2 = {x_2}^2 = x_1 \cdot x_2 = 0]]$.

More generally, the objects $D_r(n)$ represent r-jets of elements of R^{R^n}. This idea, which goes back to C. Ehresmann [40] was extended by A. Weil [111] to deal with iterated (partial) derivatives, hence the necessity to postulate the axiom 'of line type' for any Weil algebra W. For instance, the object $D_{r_1}(n_1) \times D_{r_2}(n_2)$ is not of the form $D_r(n)$ for any r, n.

An alternative to (the special cases of) Axiom J is the following [100, 61]:

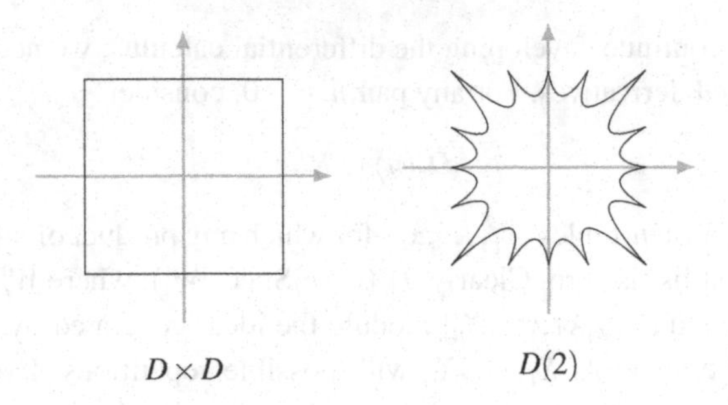

Figure 2.2 $D \times D$ versus $D(2)$

Postulate 2.5 (Postulate F) *The ring object R satisfies*

$$\forall f \in R^R \, \exists! g \in R^{R \times R} \left[\forall x, y \in R \left(f(x) - f(y) = g(x,y) \cdot (x-y) \right) \right].$$

This postulate, also called the *Reyes-Fermat Axiom*, does not use nilpotent elements of R. It is useful in certain portions of the theory. The unique g is denoted ∂f. On account of Axiom J, $\partial f(x,x) = f'(x)$. As shown by A. Kock [61], unique existence is decided on the spot, not locally. The following two results, meant to be interpreted internally, follow easily from Postulate F.

Corollary 2.6 *For all $f \in R^{R^n}$ there exist unique $g_1, \ldots, g_n \in R^{R^n \times R^n}$ such that*

(i) $\forall x, y \in R^n \; f(x) - f(y) = \sum_{i=1}^{n} g_i(x,y) \cdot (x_i - y_i)$.

(ii) $\forall x \in R^n \; \bigwedge_{i=1}^{n} g_i(x,x) = \frac{\partial f}{\partial x_i}(x)$.

Denote by $\mathfrak{M}_{n \times p}(R)$ the object of all $n \times p$-matrices with entries in R.

Corollary 2.7 *For all $f \in R^{pR^n}$ there exists a unique $g \in (\mathfrak{M}_{n \times p}(R))^{R^n \times R^n}$ such that*

(i) $\forall x, y \in R^n \; f(x) - f(y) = g(x,y) \cdot (x_i - y_i)$.

(ii) $\forall x \in R^n \; g(x,x) = D_x f$.

Axiom 2.8 **(Axiom W)** *For any Weil algebra W, the object* $\mathrm{Spec}_R(W)$ *of* $\mathscr{E}$ *is an atom in the sense of Definition 1.24.*

Remark 2.9 If W is a Weil algebra in $\mathscr{E}$, it has a unique global section, hence is well supported ($W \twoheadrightarrow 1$) and is therefore a tiny object (in fact an infinitesimal) in the sense of Definition 1.24.

For A a commutative ring with 1, denote by A^* the subobject of A consisting of its invertible elements.

Postulate 2.10 (Postulate K) (R is a field in the sense of Kock)

(K1) $\neg(1=0)$.

(K2) *For each* $n = 1,2,\ldots,$ *we have* $\neg\Big(\bigwedge_{i=1}^{n}(x_i = 0)\Big) \Rightarrow \bigvee_{i=1}^{n}(x_i \in R^*)$.

Definition 2.11 A commutative ring A with 1 is said to be a *local ring* if the following two conditions hold:

$$\neg(1=0)\,,$$

$$\forall x,y \in A\left[x+y \in A^\star \Rightarrow \left(x \in A^\star \vee y \in A^\star\right)\right].$$

Proposition 2.12 *Let* $(\mathscr{E},R)$ *satisfy Postulate K. Then the following hold:*

(i) $R^\star = \neg\{0\}$.
(ii) *R is a local ring.*

Postulate 2.13 (Postulate O) *There is a binary relation* $>$ *on* R *for which the conditions* (O1) – (O4) *(ordered) and* (O5) *(Archimedean) hold.*

(O1) $\forall x,y \in R\ \left[(x>0 \wedge y>0) \Rightarrow (x+y>0 \wedge x\cdot y>0)\right]$ *and* $1>0$.
(O2) $\forall x \in R\ \neg(x>x)$.
(O3) $\forall x,y \in R\ [x>y \Rightarrow \forall z \in R\ (x>z \vee z>y)]$.
(O4) $\forall x_1,\ldots,x_n \in R\ [\neg(\bigwedge_{i=1}^{n}(x_i=0)) \Rightarrow \bigvee_{i=1}^{n}(x_i>0 \vee x_i<0)]$.
(O5) $\forall x \in R\ \exists n \in N\ [-n<x \vee x<n]$.

Proposition 2.14 *Let* $(\mathscr{E},R)$ *be a ringed topos satisfying Postulate O (Postulate 2.13). Then the following hold:*

(i) *The relation* $>$ *is transitive, hence a strict order on R. In particular, 'intervals' can be defined as usual for* $a,b \in R$ *as*

$$(a,b) = [[x \in R \mid a<x<b]].$$

(ii) $\forall x,y \in R\ [(x>0 \wedge y>0) \Rightarrow \exists z \in R\,(z>0 \wedge z<x \wedge z<y)]$.

Proof (i) That $>$ is transitive can be shown as follows. Assuming $x>y$ and $y>z$, show that $x>z$. By (O3) we have $x>z \vee z>y$. In the first case we are done. In the second case we have $y>z \wedge z>y$. Considering the position of 0 in any of the possible situations we end up with $0>0$ which contradicts (O2).

(ii) From transitivity and strict order it follows that

$$y > x \Leftrightarrow y - x > 0.$$

Since $x > 0$, it follows that $2x - x = x > 0$, hence $2x > x$. Now, by (O3), either $y < 2x$ or $x < y$. In the first case let $z = \frac{y}{2}$. In the second case let $z = \frac{x}{2}$. □

Definition 2.15

By *SDG* it is meant the theory of a ringed topos $(\mathscr{E}, R)$ (with R a commutative ring with 1 in the topos $\mathscr{E}$) such that the following axioms and postulates are satisfied :

- Axiom **J** (jets representability),
- Axiom **W** (the jets representing objects are atoms),
- Postulate **F** (R satisfies the Reyes-Fermat condition),
- Postulate **K** (R is a field in the sense of Kock), and
- Postulate **O** (R is an Archimedean ordered ring).

The actual relevance of SDG to classical differential geometry lies in the existence of well adapted models in the following sense, which is a modification of a definition in [32, 10].

Definition 2.16

(i) A *well adapted model of ringed toposes* is a pair $(\mathscr{E}, R)$ with $\mathscr{E}$ a Grothendieck topos and R a commutative ring with 1 in it, with the additional property that, for $\mathscr{M}^\infty$ the category of smooth paracompact finite dimensional manifolds and smooth mappings, there is given an embedding (full and faithful) functor

$$\iota : \mathscr{M}^\infty \hookrightarrow \mathscr{E}$$

which preserves transversal pullbacks and the terminal object, sends the reals $\mathbb{R}$ to R and sends arbitrary open coverings in $\mathscr{M}^\infty$ to jointly epimorphic families in $\mathscr{E}$.

(ii) A *well adapted model of SDG* is a well adapted model $(\mathscr{E}, R)$ of ringed toposes which in addition is a model of SDG in the sense of Definition 2.15.

Remark 2.17 If $(\mathscr{E}, R)$ is a well adapted model of SDG, then also the tangent bundle construction is preserved. All usual notions of the differential calculus are also preserved. We refer to [61] for proofs of these and related results.

We now turn to the important concepts of tangent bundles and vector fields.

Although these notions may be defined for any object M of $\mathscr{E}$, they have particularly good properties when $M = R^m$ or, more generally, when M is an infinitesimally linear object. We postpone a discussion of this notion until the end of the section.

Just as in section 2.1 of this chapter, all notions and axioms mentioned in section 2.3 can be seen in expanded form in [61]. For this reason, and since section 2.3 is also only introductory, we do not attempt to justify certain statements that can be found therein. Alternatively, such statements may be considered as exercises for the reader.

Let $(\mathscr{E}, R)$ be a model of SDT. Let $m > 0$ and let M be an object of $\mathscr{E}$.

Definition 2.18 A *tangent vector* to M at $x \in M$ is any $t \in M^D$ such that $t(0) = x$.

As a consequence of the Kock-Lawvere axiom, this data corresponds in a unique way to $(x, v) \in M \times M$. We shall refer to v in $t = (x, v)$ as the *principal part* of t.

This leads to the subobject

$$T_x M \hookrightarrow M^D$$

of all tangent vectors to M at x, which becomes an R-module with the multiplication by scalars $r \in R$ given by

$$(r \cdot t)(d) = t(r \cdot d)$$

and the addition defined as

$$(t_1 + t_2)(d) = x + d \cdot v_1 + d \cdot v_2,$$

where v_1, v_2 are the principal parts of t_1, t_2 respectively.

If $f \colon M \longrightarrow N$ is any morphism where M and N are objects of $\mathscr{E}$, the morphism

$$M^D \xrightarrow{f^D} N^D$$

restricts to an R-linear morphism

$$(df)_x \colon T_x M \longrightarrow T_{f(x)} N.$$

Definition 2.19 We denote by

$$\pi_M \colon M^D \longrightarrow M$$

and refer to it as the *tangent bundle* of M, the morphism obtained by evaluating at $0 \in D$.

Definition 2.20 By a *vector field* on M it is meant a section $\hat{\psi}$ of the projection $\pi\colon M^D \longrightarrow M$, that is, a morphism

$$M \xrightarrow{\hat{\psi}} M^D$$

such that $\pi \circ \hat{\psi} = \mathrm{id}_M$.

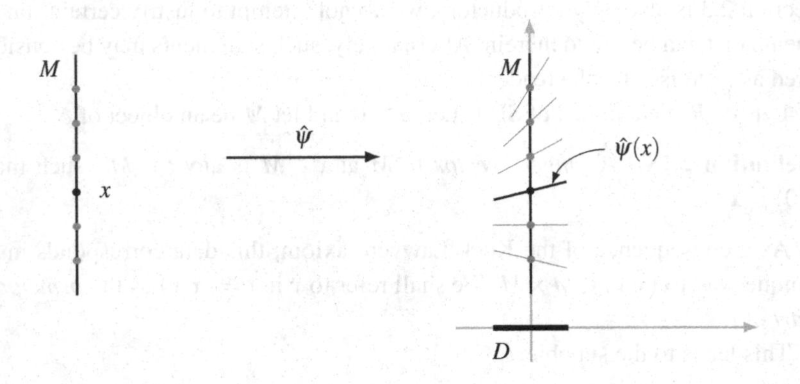

Figure 2.3 Vector field on M

Remark 2.21 By the exponential rule, available in any topos, the data for a vector field on M is equivalently given by a morphism

$$M \times D \xrightarrow{\psi} M$$

such that $\forall x \in M\ (\psi(x,0) = x)$.

Lemma 2.22 *Let ψ be a vector field on M. Then, ψ is an 'infinitesimal flow' relative to D, or a D-flow. This means that the following equation is satisfied:*

$$\forall x \in M\, \forall d_1, d_2 \in D\, [d_1 + d_2 \in D \Rightarrow \psi(x, d_1 + d_2) = \psi(\psi(x, d_1), d_2)]\,.$$

Proof Note that $d_1 + d_2 \in D$ if and only if $(d_1, d_2) \in D(2)$, so that, for a given $x \in M$, both sides of the equation are morphisms $D(2) \longrightarrow M$. They will be equal provided they have the same partial derivatives at 0. This is clear as they coincide on the axes, that is,

$$\psi(x, 0 + d_2) = \psi(x, d_2) = \psi(\psi(x, 0), d_2)$$

and

$$\psi(x, d_1 + 0) = \psi(x, d_1) = \psi(\psi(x, d_1), 0)\,.$$

□

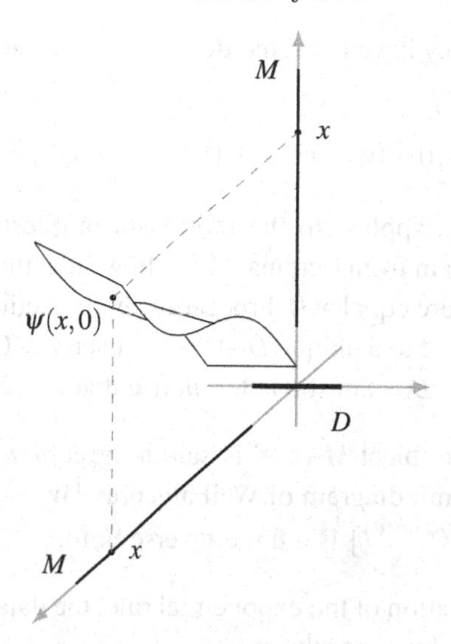

Figure 2.4 Infinitesimal flow

Remark 2.23 Lemma 2.22 justifies the terminology 'infinitesimal flow' since a global flow on $M = R^m$ would be a morphism

$$\bar{\psi}\colon M \times R \longrightarrow M$$

such that for any $x \in M$ and any $r_1, r_2 \in R$,

$$\bar{\psi}(\bar{\psi}(x, r_1), r_2) = \bar{\psi}(x, r_1 + r_2)$$

as well as $\bar{\psi}(x, 0) = x$. Thus, almost by definition, a vector field on $M = R^m$ is already integrated into D. In what follows it will be shown that for $M = R^m$ in particular, vector fields can always be integrated into D_∞-flows. In other words, it is always possible to integrate all differential equations (when $M = R^m$ but as we shall see also in more general cases) by formal power series.

Proposition 2.24 *Let $(\mathscr{E}, R)$ be a model of SDT. Let $m > 0$ and $M = R^m$. For any vector field*

$$\psi\colon M \times D \longrightarrow M,$$

the following holds for any $r > 0$:

$$\forall x \in M \ \forall (d_1, d_2) \in D_r(2) \ \big[\psi(\psi(x, d_1), d_2) = \psi(x, d_1 + d_2)\big].$$

Proof We begin with the case $n = 2$. For $(d_1, d_2) \in D_2(2)$, both sides are morphisms $D_2(2) \longrightarrow M$, that is, they are 2-jets in two variables, hence they are equal provided they have the same derivatives of order 2 at 0. We have, by Lemma 2.22, that

$$\psi(x, 0+0) = \psi(\psi(x,0),0) = \psi(x,0) = x.$$

Thus, $\frac{\partial^2}{\partial x^2}$ and $\frac{\partial^2}{\partial y^2}$, applied to the morphism in question, are equal when evaluated at 0. But again from Lemma 2.22 follows that they agree on $D \times D$, so also $\frac{\partial}{\partial x}$, $\frac{\partial}{\partial y}$ and $\frac{\partial^2}{\partial x \partial y}$ are equal at 0. Proceeding by induction we derive that any D-flow can be extended to a unique D_r-flow for each $r > 0$, hence has a unique extension to a flow on D_∞. For the latter, notice that $D_\infty(2) = D_\infty \times D_\infty$. □

Definition 2.25 An object M of $\mathscr{E}$ is said to be *infinitesimally linear* if for every finite inverse limit diagram of Weil algebras $\{W \longrightarrow W_i\}$, we have that $\{M^{Spec_R(W_i)} \longrightarrow M^{Spec_R(W)}\}$ is a finite inverse limit.

By a further application of the exponential rule, the data of a vector field can equivalently be given by a morphism

$$D \xrightarrow{\check{\psi}} M^M$$

such that $\check{\psi}(0) = \mathrm{id}_M$, in other words, an 'infinitesimal path' in the 'space' of all deformations of the identity.

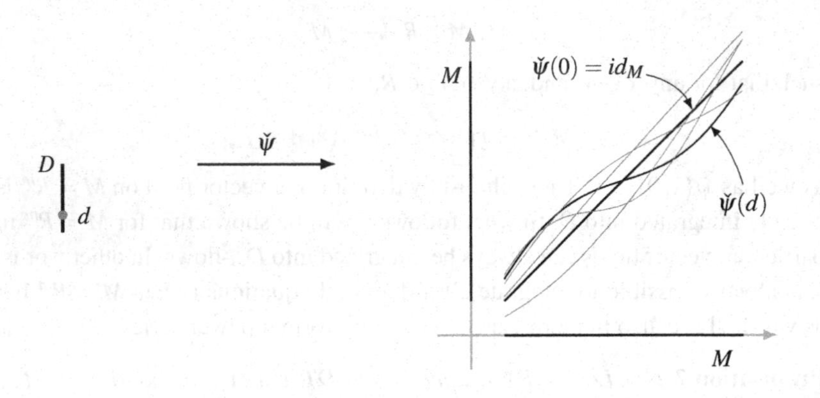

Figure 2.5 Infinitesimal deformations of id_M

For an infinitesimally linear object M, such infinitesimal deformations of the identity are bijective morphisms or permutations of the elements of M. In particular, for any vector field ψ on M, we have

$$\forall xp \in M \, \forall d \in D[\psi(\psi(x,d),-d) = x]$$

and, in particular, again, each infinitesimal deformation $\breve{\psi}(d)\colon M \longrightarrow M$ is invertible with inverse $\breve{\psi}(-d)$.

Proposition 2.26 *The class of infinitesimally linear objects in a model $(\mathscr{E},R)$ of SDG contains R, and is closed under finite limits and exponentials by arbitrary objects.*

Proof That R, assumed here to be a Q-algebra, is infinitesimally linear, is a consequence of the following Lemma.

Lemma 2.27 (Kock-Lavendhomme) *If $\{W \longrightarrow W_i\}$ is any finite inverse limit of Weil algebras, then for any Q-algebra A,*

$$\{A \otimes W \longrightarrow A \otimes W_i\}$$

is also a finite inverse limit.

That the class of infinitesimally linear objects of $\mathscr{E}$ is closed under finite limits and exponentials by arbitrary objects is immediate. □

Proposition 2.28 *Vector fields on an arbitrary infinitesimally linear object M can be integrated into D_∞.*

The generality afforded by considering all Weil algebras in the definition of infinitesimal linearity is not always needed. Particular cases of this notion that we will need the most are the following ones and their combinations.

Proposition 2.29 *Let M be an infinitesimally linear object of $\mathscr{E}$, where $(\mathscr{E},R)$ is a model of Axiom J.*

(i) *Consider in $\mathscr{E}$ the diagram*

$$D \times D \underset{\sigma}{\overset{\mathrm{id}}{\rightrightarrows}} D \times D \xrightarrow{+} D_2,$$

with $\sigma(d_1,d_2) = (d_2,d_1)$. Then for every $\tau\colon D \times D \longrightarrow M$ with $\tau(t,d) = \tau(d,t)$ for all $(t,d) \in D \times D$, there exists a unique $\psi\colon D_2 \longrightarrow M$ such that $\psi(t+d) = \tau(t,d)$.

(ii) *Consider the diagram*

$$\begin{array}{ccc} D & \longrightarrow & D \times D_0 \\ \downarrow & & \downarrow \\ D_0 \times D & \longrightarrow & D(2) \end{array}$$

Then for any pair of morphisms $\tau_i\colon D \longrightarrow M$ *(*$i = 1,2$*) with* $\tau_1(0) = \tau_2(0)$ *there exists a unique morphism*

$$\ell\colon D(2) \longrightarrow M$$

such that $\ell \circ \mathrm{incl}_i = \tau_i$ *for* $i = 1,2$.

Proof (i) This can be stated in terms of finite limits of Weil algebras. Indeed, to the equalizer

$$Q[\theta] \xrightarrow{\varphi} Q[\varepsilon,\delta] \underset{\sigma}{\overset{\mathrm{id}}{\rightrightarrows}} Q[\varepsilon,\delta],$$

where $\theta^3 = 0$, $\varepsilon^2 = \delta^2 = 0$, $\varphi(\theta) = \varepsilon + \delta$, $\sigma(\varepsilon) = \delta$ and $\sigma(\delta) = \varepsilon$, corresponds via $\mathrm{Spec}_R(-)$ the diagram

$$D \times D \underset{\sigma}{\overset{\mathrm{id}}{\rightrightarrows}} D \times D \xrightarrow{+} D_2$$

in $\mathscr{E}$, with $\sigma(d_1,d_2) = (d_2,d_1)$. In turn, taking the exponential with M yields

$$M^{D_2} \longrightarrow M^{D\times D} \rightrightarrows M^{D\times D}.$$

To say that this is an equalizer means exactly that for every $\tau\colon D\times D \longrightarrow M$ with $\tau(t,d) = \tau(d,t)$ for all $(t,d) \in D\times D$, there is a unique $\psi\colon D_2 \longrightarrow M$ such that $\psi(t+d) = \tau(t,d)$.

(ii) This too can be expressed in terms of finite limits of Weil algebras. Consider the diagram

$$\begin{array}{ccc} Q[\varepsilon,\delta] & \longrightarrow & Q[\alpha,\beta] \\ \downarrow & & \downarrow \\ Q[\eta,\mu] & \longrightarrow & Q \end{array}$$

of Weil algebras, where $\varepsilon^2 = \delta^2 = 0$, $\varepsilon\cdot\delta = 0$, $\alpha = 0$, $\beta^2 = 0$, $\eta^2 = 0$ and $\mu = 0$, which is easily seen to be a pullback. This diagram corresponds via $\mathrm{Spec}_R(-)$ to a diagram

$$\begin{array}{ccc} D & \longrightarrow & D \times D_0 \\ \downarrow & & \downarrow \\ D_0 \times D & \longrightarrow & D(2) \end{array}$$

and in turn, by Axiom J, to the pullback diagram

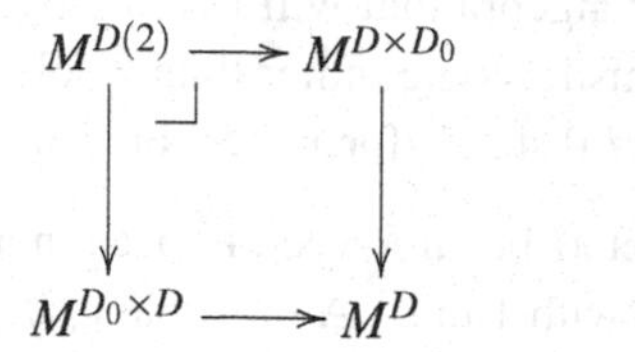

That this is a pullback diagram translates into the statement that for any pair of morphisms $\tau_i\colon D \longrightarrow M$ $(i=1,2)$ with $\tau_1(0)=\tau_2(0)$ there exists a unique morphism

$$\ell\colon D(2) \longrightarrow M$$

such that $\ell \circ \mathrm{incl}_i = \tau_i$ for $i=1,2$. □

Corollary 2.30 *Let M be an infinitesimally linear object in $\mathscr{E}$, where $(\mathscr{E},R)$ is a model for axiom J. Then for each $x\in M$, T_xM is an R-module.*

Proof The external multiplication can be defined, without any assumption on M, as follows

$$(r\cdot\tau)(d)=\tau(r\cdot d),$$

for any $\tau\in T_xM$, $r\in R$, and $d\in D$.

As for the addition, given $\tau_1,\tau_2\in M^D$ such that $\tau_1(0)=\tau_2(0)=x$, we define

$$(\tau_1+\tau_2)(d)=\ell(d,d),$$

where $\ell\in M^{D(2)}$ is the unique map obtained in (2) so that $\ell\circ\mathrm{inc}_i=\tau_i$ for $i=1,2$. □

Remark 2.31 The name 'infinitesimal linear' was originally used [61] to refer to a more general form of condition on conclusion 2 in Proposition 2.29 where n is arbitrary rather than $n=2$. In Definition 2.25 we use infinitesimal linearity in the (strong) sense of Bergeron [10] (called strong infinitesimal linearity in [61]). In the future we will mostly use Axiom J for M infinitesimally linear and objects $D_r(n)$. That M is infinitesimally linear implies in this case that r-jets of functions of n variables with values on M behave as if they were the list of its partial derivatives up to order r at zero.

2.2 Linear Algebra in SDG

We start by stating some notions and results from linear algebra that will be needed in what follows. In the context of synthetic differential geometry, all notions are defined for a topos $\mathscr{E}$ with a ring object R in it. This means that the notions of linear algebra that will be employed are subject to the rules of Heyting (or intuitionistic) logic rather than Boolean (or classical) logic.

In what follows, $x\#0$ stands for $x \in R^*$ or 'x invertible'.

Definition 2.32 Let M be an R-vector space in a topos $\mathscr{E}$, where R is a commutative ring object with 1 in $\mathscr{E}$. An n-tuple $v_1, \ldots, v_n \in M$ is said to be

(i) *linearly free* if

$$\forall \lambda_1, \ldots, \lambda_n \in R \left[\bigvee_{i=1}^{n} (\lambda_i \# 0) \Rightarrow \neg \Big(\sum_{i=1}^{n} \lambda_i v_i = 0 \Big) \right]$$

holds in $\mathscr{E}$, and

(ii) *linearly independent* if

$$\forall \lambda_1, \ldots, \lambda_n \in R \left[\sum_{i=1}^{n} \lambda_i v_i = 0 \Rightarrow \bigwedge_{i=1}^{n} (\lambda_i = 0) \right]$$

holds in $\mathscr{E}$.

Remark 2.33 It is worth to point out that in the setting of Definition 2.32 the definitions of linearly free and linearly independent are not equivalent, the latter being stronger in any ringed topos $(\mathscr{E}, R)$ in which $\neg(1 = 0)$. Indeed, given linearly independent vectors $v_1, \ldots, v_n \in M$, from the definition it follows that

$$\forall \lambda_1, \ldots, \lambda_n \in R \left[\neg \Big(\bigwedge_{i=1}^{n} (\lambda_i = 0) \Big) \Rightarrow \neg \Big(\sum_{i=1}^{n} \lambda_i v_i = 0 \Big) \right].$$

From $\neg(1 = 0)$ we get that $x\#0 \Rightarrow \neg(x = 0)$ and therefore, $\forall \lambda_1, \ldots, \lambda_n \in R$,

$$\bigvee_{i=1}^{n} (\lambda_i \# 0) \Rightarrow \bigvee_{i=1}^{n} (\neg(\lambda_i = 0)) \Rightarrow \neg \Big(\bigwedge_{i=1}^{n} (\lambda_i = 0) \Big) \Rightarrow \neg \Big(\sum_{i=1}^{n} \lambda_i v_i = 0 \Big).$$

We could have the implication in the reverse direction in all generality, if our apartness relation were strict, in the sense that $\neg(x\#0)$ iff $(x = 0)$, which is not the case. However if the ringed topos $(\mathscr{E}, R)$ satisfies Postulate K, then the situation is different, as noted in the next result.

Recall that, for a ringed topos $(\mathscr{E}, R)$, if R is a field in the sense of Kock, we have

$$\neg(x = 0) \Leftrightarrow (x\#0).$$

It is desirable, however, to refer to the property of an element x being apart from 0 $(x\#0)$ in certain places, whereas in others it is better suited to state that x is not equal to 0 $(\neg(x = 0))$.

Proposition 2.34 *Let $(\mathscr{E}, R)$ be a ringed topos on which Postulate K holds and let $v_1, \ldots, v_n \in R^k$ with $n \leq k$. Then, the n-tuple $v_1, \ldots, v_n \in R^k$ is linearly free if and only if it is linearly independent.*

Proof The sufficiency has already been observed earlier. We then show the necessity. Assume that $v_1, \ldots, v_n \in R^k$ is linearly free, where $n \leq k$. Let $A \in \mathfrak{M}_{k \times n}(R)$ be the matrix whose rows are formed with the coordinates of the v_i's. We will find a minor of order n that is invertible. Indeed, each vector v_i, as every subset of linearly free vectors, is linearly free, and therefore v_1 must be different from $\bar{0} \in R^k$, i.e.

$$\neg\left(\bigwedge_{j=1}^{k} (a_{1j} = 0)\right), \quad \text{and by Postulate K,} \quad \bigvee_{j=1}^{k} (a_{1j} \# 0).$$

Assume by simplicity that $a_{11} \# 0$, and use it as pivot to sweep down all the entries below in the matrix A by means of elementary row operations to get the matrix

$$\begin{pmatrix} a_{11} & a_{12} & \cdots & a_{1k} \\ 0 & b_{22} & \cdots & b_{2k} \\ \vdots & \vdots & & \vdots \\ 0 & b_{n2} & \cdots & b_{nk} \end{pmatrix}.$$

Clearly these manipulations do not affect the linear freedom of the rows nor their linear independence, as it is easy to verify that for any R-module M, given $u, v \in M$ and $\lambda \in R$, the vectors u and v are linearly free (resp. independent) if and only if the vectors u and $\lambda u + v$ are linearly free (resp. independent). Therefore, as before $(0, b_{22}, \ldots, b_{2k})$ is linearly free and, assuming that $b_{22} \# 0$ we sweep all the entries below and keep going to end up with an invertible n-minor. Now, as it was proved in [59], these n rows are linearly independent and so are the original vectors, and we are done. □

In what follows we shall cast our results in terms of the notion of n linearly free vectors of R^k for $n \leq k$ while availing ourselves of all results already shown in [59] for linear independence, in view of Proposition 2.34. The main result in this connection is the one relating the notions of row rank, column rank, and invertibility of the determinant which we quote below.

Proposition 2.35 *[59] Assume that $(\mathscr{E}, R)$ is a model of SDG.*

(i) *Let $X \in \mathfrak{M}_{p \times n}(R)$. Then,* $\mathrm{rowRank}(X) \geq r \Leftrightarrow \mathrm{columnRank}(X) \geq r$. *(One writes* $\mathrm{Rank}(X) = r$ *for* $\mathrm{rowRank}(X) \geq r$ *or* $\mathrm{columnRank}(X) \geq r$ *when $r = p$ or $r = n$ respectively.)*

(ii) *Let $X \in \mathfrak{M}_{p\times n}(R)$ be such that* $\mathrm{Rank}(X) = p$. *Then, locally, X has a right inverse.*

Definition 2.36 Let M be an R-vector space in $\mathscr{E}$, and N an R-subspace of M.

(i) We say that *the rank of* N *is at least n*

$$\mathrm{Rank}(N) \geq n$$

if there exist vectors $v_1, \ldots, v_n \in N$ which are linearly free.

(ii) We say that *the independence of* N *is at most n*

$$\mathrm{indep}(N) \leq n$$

if there exist n vectors $v_1, \ldots, v_n \in N$ such that $N \subset \neg\neg\langle v_1, \ldots, v_n\rangle$.

(iii) We say that *the dimension of* N *equals n*

$$\dim(N) = n$$

if $\mathrm{rank}(N) \geq n$ and $\mathrm{indep}(N) \leq n$.

Exercise 2.37 Prove that if $\mathrm{indep}(N) \leq n$ then it cannot be the case that $\mathrm{rank}(N) \geq n+1$. In particular, given a chain of subspaces

$$V_0 \subset V_1 \subset \cdots \subset V_n = N$$

the existence of $v_0, v_1, \ldots, v_n$ with $\neg(v_0 = 0)$, and $v_i \in V_i \backslash V_{i-1}$ for $i = 1, \ldots, n$ is contradictory with $\mathrm{indep}(N) \leq n$. Indeed, the assumption would imply the existence of $n+1$ linearly free vectors in N.

Lemma 2.38 (Nakayama's lemma) *Let A be a ring with unit, and let $\mathfrak{m} \subset A$ be an ideal such that for each $a \in \mathfrak{m}$, $1+a$ is invertible in A. Let M be a finitely generated A-module and let N be an A-submodule of M, $N \subset M$. If $\mathfrak{m}M + N = M$, then $N = M$.*

Proof From $M = \langle m_1, \ldots, m_p\rangle$ and $mM + N = M$ one gets $m_i = n_i + \sum_{j=1}^{p} a_{ij} m_j$ where $a_{ij} \in \mathfrak{m}$ for $i = 1, \ldots, p$ and $n_i \in N$. Now, the matrix given by

$$\begin{pmatrix} 1-a_{11} & -a_{12} & \cdots & -a_{1p} \\ -a_{21} & 1-a_{22} & \cdots & -a_{2p} \\ \vdots & \vdots & \ddots & \vdots \\ -a_{p1} & -a_{p2} & \cdots & 1-a_{pp} \end{pmatrix}$$

is invertible since its determinant is of the form $1+a$ with $a \in \mathfrak{m}$. Therefore, for the column $(p \times 1)$-matrices $X = (m_1 \cdots m_p)^t$ and $Y = (n_1 \cdots n_p)^t$

$$\begin{pmatrix} 1-a_{11} & -a_{12} & \dots & -a_{1p} \\ -a_{21} & 1-a_{22} & \dots & -a_{2p} \\ \vdots & \vdots & \ddots & \vdots \\ -a_{p1} & -a_{p2} & \dots & 1-a_{pp} \end{pmatrix} \begin{pmatrix} m_1 \\ m_2 \\ \vdots \\ m_p \end{pmatrix} = \begin{pmatrix} n_1 \\ n_2 \\ \vdots \\ n_p \end{pmatrix}$$

can be solved. Therefore $M \subset N$ and so $M = N$. □

Corollary 2.39 *Let A be a unitary ring and let $\mathfrak{a}$ be an ideal of A such that for each $a \in \mathfrak{a}$, $1+a$ is invertible. Let M be a finitely generated A-module. If $\mathfrak{a}M = M$ then $M = \{0\}$.*

Theorem 2.40 *Let A be an R-algebra and $\mathfrak{m} \subset A$ an ideal such that $1+a$ is invertible for each $a \in \mathfrak{m}$. Let V be an A-submodule with* $\text{indep}(V) < \infty$ *and $M \subset V$ an A-submodule of V such that* $\text{indep}(V/M) \leq n$. *Then, $\mathfrak{m}^n V \subset \neg\neg M$.*

Proof Under the assumption that in the R-vector space V/M there are no more than n free vectors, we must show that $\mathfrak{m}^n V \subset \neg\neg M$. For this, it is enough to show the validity of

$$\neg[\exists v \in \mathfrak{m}^n V/M |\ \neg(v = 0)].$$

Consider the chain of A-submodules

$$\mathfrak{m}^n V/M \subset \mathfrak{m}^{n-1} V/M \subset \cdots \subset V/M$$

where $\mathfrak{m}^i V/M$ denotes $\mathfrak{m}^i V/M \cap \mathfrak{m}^i V$. For each $i = 1, 2 \ldots n$,

$$v \in \mathfrak{m}^i V/M = \mathfrak{m}(\mathfrak{m}^{i-1} V/M).$$

In addition, $v \in \mathfrak{m}^{i-1} V/M$ and since $\neg(v = 0)$, then

$$\neg(\mathfrak{m}^i V/M = \mathfrak{m}^{i-1} V/M)$$

for $i = 1, 2, \ldots, n$ by Corollary 2.39 which is a consequence of Lemma 2.38 (Nakayama's lemma). Therefore

$$\neg\neg[\exists v_{i-1} \in (\mathfrak{m}^{i-1} V/M) \setminus (\mathfrak{m}^i V/M)]$$

for $i = 1, 2, \ldots n$, which contradicts Exercise 2.37. □

We shall need the following.

Lemma 2.41 *Let X and Y be R-vector spaces and $h \colon X \twoheadrightarrow Y$ an epimorphism. If* $\text{indep}(X) \leq n$, *then also* $\text{indep}(Y) \leq n$.

Proof If the independence of X is at most n, then there exist $v_1,\ldots,v_n \in X$ such that $X \subset \neg\neg\langle v_1,\ldots,v_n\rangle$. Then, $Y = \mathrm{Im}(h) \subset \neg\neg\langle h(v_1),\ldots,h(v_n)\rangle$ and therefore $\mathrm{indep}(Y) \leq n$. □

PART TWO

TOPICS IN SDG

In this second part we illustrate the principles of synthetic differential geometry and topology in two distinct areas. The first example is a theory of connections and sprays, where we show that—unlike the classical situation—the passage from connections to geodesic sprays need not involve integration, except in infinitesimal form. Moreover, the validity of the Ambrose-Palais-Singer theorem within SDG extends well beyond the classical one. In our second example we show how in SDG one can develop a calculus of variations 'without variations', except for those of an infinitesimal nature. Once again, the range of applications of the calculus of variations within SDG extends beyond the classical one. Indeed, in both examples, we work with infinitesimally linear objects—a class closed under finite limits, exponentiation, and étale descent. The existence of well adapted models of SDG guarantees that those theories developed in its context are indeed relevant to the corresponding classical theories.

3
The Ambrose-Palais-Singer Theorem in SDG

A theorem of W. Ambrose, R.S. Palais and I.M. Singer [1] establishes a bijective correspondence between torsion-free affine connections on a finite dimensional smooth manifold M and sprays on M. The notions contained in the statement of this theorem are all expressible in SDG using infinitesimals, yet the classical proof employs local (not infinitesimal) concepts. Our goal is to show that there is a simple generalization of the classical proof of the Ambrose-Palais-Singer theorem on connections and sprays to the class of infinitesimally linear objects based on [28]. To this end, we remark [22] that the local notions employed in it ('iterated tangent bundle', 'existence of the exponential map') are themselves consequences of infinitesimal linearity [65, 63]. This is an instance of a method often employed in SDG, which is to proceed from the local to the infinitesimal and in the process obtain a result that is considerably more general than its classical counterpart. A direct proof for infinitesimally linear objects, however, has alternatively been given [22, 89]. In this chapter we shall assume that the ring R in a topos $\mathscr{E}$ is a Q-algebra and that the object M of $\mathscr{E}$ is infinitesimally linear. All notions and statements from synthetic differential geometry that are employed in the present chapter have been reviewed in the first part of this book.

3.1 Connections and Sprays

Let $(\mathscr{E}, R)$ be a model of SDG (in the sense of Definition 2.15) and let M be an infinitesimally linear object (see Definition 2.25) of $\mathscr{E}$. Among the consequences of infinitesimal linearity of M is the *euclideanness* of the tangent bundle of M. This means that the tangent bundle

$$\pi_M : M^D \longrightarrow M$$

is a trivial bundle (or that M is 'parallelizable') in the sense that there is a Euclidean R-module V and an isomorphism $\rho\colon M^D \simeq M \times V$ such that the following diagram commutes.

$$\begin{array}{ccc} M^D & \xrightarrow{\rho} & M \times V \\ & \pi_M \searrow \quad \swarrow \text{proj}_1 & \\ & M & \end{array}$$

We shall now abstract the above set-up as follows. Let $p\colon E \longrightarrow M$ be a vector bundle, that is, an R-module in the category $\mathscr{E}/M$, with E and M both infinitesimally linear.

Remark 3.1 The basic example of a vector bundle is the tangent bundle, but the more general situation will be better suited to express parallel transport in that it notationally distinguishes between the vector being transported and the vector which effects the transport.

On E^D there are two linear structures. Since E is infinitesimally linear, there is the tangential addition on the tangent bundle $\pi_E\colon E^D \longrightarrow E$. We shall denote this addition by $\oplus$. Corresponding to it is the scalar multiplication given by

$$(\lambda \odot f)(d) = f(\lambda \cdot d)\,.$$

E^D is also equipped with $p^D\colon E^D \longrightarrow M^D$. For $f, g \in M^D$ in the same fibre (i.e. $p^D(f) = p^D(g)$, or $p \circ f = p \circ g$), there is defined an addition

$$(f+g)(d) = f(d) + g(d)$$

which, together with

$$(\lambda \cdot f)(d) = \lambda \cdot f(d),$$

gives an R-module structure on each fibre.

We have the morphisms

$$M^D \times_M E \overset{\text{proj}_1}{\longrightarrow} M^D \overset{\pi_M}{\longrightarrow} M$$

and

$$M^D \times_M E \overset{\text{proj}_2}{\longrightarrow} E \overset{p}{\longrightarrow} M$$

which respectively define $\oplus$ and $+$, giving in turn two linear structures on $M^D \times_M E$ in the obvious fashion.

The object part E^D of the tangent bundle $\pi_E : E^D \longrightarrow E$ is also equipped with a map $p^D : E^D \longrightarrow M^D$ and the morphism

$$K = \langle p^D, \pi_D \rangle : E^D \longrightarrow M^D \times_M E$$

is linear with respect to the two structures, as it is easy to verify [60].

Definition 3.2 A *connection* on $p : E \longrightarrow M$ is a morphism

$$\nabla : M^D \times_M E \longrightarrow E^D$$

which is a splitting of K and which is linear with respect to the two linear structures.

Remark 3.3 Consider the tangent bundle $\pi_M : M^D \longrightarrow M$ of M, for M infinitesimally linear. In this special case, a connection is equivalently given by a rule which, given $(t, v) \in M^D \times V$, allows for v to be transported in parallel fashion along an infinitesimal portion of the curve with velocity vector t. This can be justified by means of the euclideanness property, whereby there is an isomorphism $\rho : M^D \cong M \times V$ over M, where V is a Euclidean R-module. This property is a consequence of infinitesimal linearity. By means of parallel transport it is possible to compare velocity vectors attached to different (yet nearby) positions along a curve.

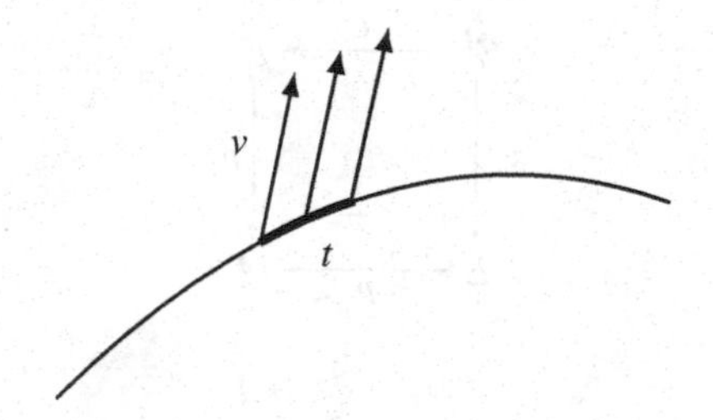

Figure 3.1 Parallel transport

This transport must therefore be required to be a linear map between the tangent spaces involved, that is, with respect to $\oplus$ and $\odot$. One reason for the need to compare nearby velocity vectors is in order to compute acceleration in terms of small changes in them. Given a motion $\xi(t)$ on M, we may always consider the iterated vector field $\xi''(t)$ of the velocity field $\xi'(t)$ but, in order to interpret it as acceleration, it is necessary to have a rule which allows for the reduction of second-order data to first-order data. In other words, what is needed for this is a map

$$C : (M^D)^D \longrightarrow M^D$$

satisfying a certain property.

Let

$$v : M^D \rightarrowtail (M^D)^D$$

be the map which identifies the fibre of the tangent bundle $\pi_M : M^D \longrightarrow M$ above $m \in M$ with the tangent space to this fibre at 0_m and which is given by the rule

$$v \mapsto [d \mapsto d \cdot v].$$

It is natural to require that this lifting of first-order data to second-order data in a trivial fashion give back the original data when C is applied. Moreover, some linearity assumptions are in order. We do this next in more generality.

Definition 3.4 A *connection map* on a vector bundle $p : E \longrightarrow M$, where E and M are both infinitesimally linear, is a map

$$C : E^D \longrightarrow E$$

which fits into the commutative squares [1]

$$\begin{array}{ccc} E^D & \xrightarrow{C} & E \\ \Big\downarrow{\scriptstyle \pi_E} & & \Big\downarrow{\scriptstyle p} \\ E & \xrightarrow[p]{} & M \end{array}$$

and

$$\begin{array}{ccc} E^D & \xrightarrow{C} & E \\ \Big\downarrow{\scriptstyle p^D} & & \Big\downarrow{\scriptstyle p} \\ M^D & \xrightarrow[\pi_M]{} & M \end{array} ,$$

is such that $C \circ v = \mathrm{id}_E$ and is linear with respect to the two structures $\oplus$, $+$ on E^D.

Definition 3.5 A vector bundle $p : E \longrightarrow M$ is said to have the *short path*

[1] On account of the naturality of the 'base point maps' π, the commutativity on any one of the two squares implies the commutativity of the other.

lifting property if, given any object X and a commutative square

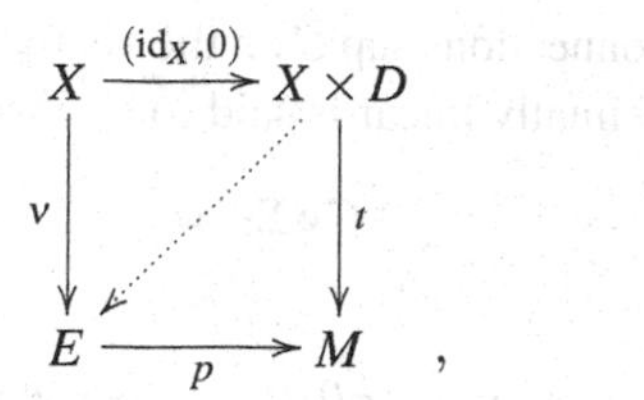

a diagonal fill-in exists as shown (but it is not required to be unique).

Proposition 3.6 *Let $p\colon E \longrightarrow M$ be a Euclidean R-module in $\mathscr{E}/M$, with E and M both infinitesimally linear objects. Assume that $p\colon E \longrightarrow M$ has the short path lifting property. Then the data for a connection on $p\colon E \longrightarrow M$ is equivalent to the data for a connection map on $p\colon E \longrightarrow M$.*

Proof Consider the 'horizontal' map

$$M^D \times_M E \xrightarrow{H} E^D$$

which is the kernel of K. It can be verified that

$$K \circ (\mathrm{id} \ominus (\nabla \circ K)) = 0$$

and so, a unique morphism $C\colon E^D \longrightarrow E$ exists such that the diagram

$$\begin{array}{ccc} E^D & \xrightarrow{\mathrm{id}\ominus(\nabla\circ K)} & E^D \\ & {\scriptstyle \langle p^D, C\rangle} \searrow \quad \nearrow {\scriptstyle H} & \\ & M^D \times_M E & \end{array}$$

commutes. Since H is a monomorphism, it follows from

$$H \circ \langle p^D, C\rangle \circ \nabla = (\mathrm{id} \ominus (\nabla \circ K)) \circ \nabla = \nabla \ominus (\nabla \circ K \circ \nabla) = 0$$

that

$$C \circ \nabla = 0$$

as desired. The linearity properties are easily checked [28]. □

Exercise 3.7 Prove that if a connection exists on a vector bundle $p\colon E \longrightarrow M$ then $p\colon E \longrightarrow M$ has the short path lifting property. This is always the case for the tangent bundle $\pi_M\colon M^D \longrightarrow M$ where M is infinitesimally linear, so in particular M is parallelizable. Indeed, a splitting for

$$M \times V \times V \times V \xrightarrow{K} M \times V \times V,$$

where $K(m,u,v,w) = (m,u,v)$, is simply given by the rule $(m,u,v) \mapsto (m,u,v,0)$.

Definition 3.8 A connection map C on the vector bundle $p\colon E \longrightarrow M$ where E and M are infinitesimally linear is said to be *torsion free* if it satisfies

$$C \circ \Sigma_E = C$$

where

$$\Sigma_E\colon (E^D)^D \longrightarrow (E^D)^D$$

is the isomorphism given as the composite

$$(E^D)^D \xrightarrow{\varphi} E^{D\times D} \xrightarrow{E^\tau} E^{D\times D} \xrightarrow{\varphi^{-1}} (E^D)^D$$

where τ is the twist map and φ the canonical isomorphism.

Exercise 3.9 A connection ∇ on the tangent bundle $\pi_M\colon M^D \longrightarrow M$ of M is *torsion free* if and only if it satisfies

$$\nabla(v_1,v_2)(d_2,d_1) = \nabla(v_2,v_1)(d_1,d_2)$$

for all $(v_1,v_2) \in M^D \times_M M^D \longrightarrow (M^D)^D$ and $d_1,d_2 \in D$.

In the next sections we shall assume that $p\colon E \longrightarrow M$ is a vector bundle with E and M both infinitesimally linear and that it satisfies the short path lifting property. We shall therefore use the notions of a connection and of a connection map on $p\colon E \longrightarrow M$ interchangeably on account of Proposition 3.6.

A connection map C on a vector bundle $p\colon E \longrightarrow M$ leads to covariant differentiation of a vector field along a curve in a way that generalizes the definition of the formal derivative of a map $f\colon R \longrightarrow V$ when V is a Euclidean R-module. We recall the latter.

Denoting by $\gamma\colon V^D \longrightarrow V$ the principal part, define

$$\frac{df}{dt}\colon R \longrightarrow V$$

as the composite

$$R \xrightarrow{\dagger} R^D \xrightarrow{f^D} V^D \xrightarrow{\gamma} V$$

or, defining $R \xrightarrow{f'} V^D$ as the composite $R \xrightarrow{\dagger} R^D \xrightarrow{f^D} V^D$,

$$\frac{df}{dt}\colon R \longrightarrow V = R \xrightarrow{f'} V^D \xrightarrow{\gamma} V\,.$$

Assume now that $C\colon E^D \longrightarrow E$ is a connection map on $p\colon E \longrightarrow M$,

$a\colon R \longrightarrow M$ a 'curve' and $X\colon R \longrightarrow E$ a vector field above a. As above, we can always form

$$R \xrightarrow{X'} E^D = R \xrightarrow{\dagger} R^D \xrightarrow{X^D} E^D.$$

It follows that $X'\colon R \longrightarrow E^D$ so defined is a vector field on $p^D\colon E^D \longrightarrow M^D$ above $a'\colon R \longrightarrow M^D$. Define

$$R \xrightarrow{\frac{DX}{dt}} E = R \xrightarrow{X'} E^D \xrightarrow{C} E$$

where it is implicit that the so defined (and easily verified to be a) vector field $\frac{DX}{dt}\colon R \longrightarrow E$ on $p\colon E \longrightarrow M$ above a depends on C.

Remark 3.10 This is indeed a generalization of the derivative of $f\colon R \longrightarrow V$ regarding $V \longrightarrow 1$ as a vector bundle with one fibre. Since V is a Euclidean R-module, the principal part $\gamma\colon V^D \longrightarrow V$ is a connection map on $V \longrightarrow 1$. Any map $f\colon R \longrightarrow V$ may be regarded as a vector field on V along the unique curve $R \longrightarrow 1$.

From now on we shall focus on the tangent bundle $\pi_M\colon M^D \longrightarrow M$ of an infinitesimally linear object M of a topos $\mathscr{E}$, where $(\mathscr{E}, R)$ is a model of SDG. Given any curve $a\colon R \longrightarrow M$ of M, there is a canonical vector field over a on the tangent bundle $\pi_M\colon M^D \longrightarrow M$, to wit $a'\colon R \longrightarrow M^D$. Hence, if C is a connection on $\pi_M\colon M^D \longrightarrow M$, the *covariant derivative* $\frac{Da'}{dt}\colon R \longrightarrow M^D$ is defined and is a vector field over a.

Definition 3.11 A curve $a\colon R \longrightarrow M$ is a *geodesic* with respect to a connection map C on the tangent bundle $\pi_M\colon M^D \longrightarrow M$ if $\frac{Da'}{dt} = 0$.

Exercise 3.12 Show that an alternative notion of geodesic can be given in terms of the connection map C itself, equivalently, in terms of its associated connection ∇, as follows. A curve $X\colon R \longrightarrow M^D$ on M^D is said to be ∇-parallel [60] if

$$X' = \nabla((\pi_M \circ X)', X)$$

and $a\colon R \longrightarrow M$ is a geodesic for ∇ if a' is ∇-parallel. Similarly, a is a geodesic for C if $C \circ a'' = 0$.

Definition 3.13 A *spray* on the tangent bundle of M is here defined as a morphism

$$S\colon M^D \longrightarrow (M^D)^D$$

satisfying

(i) $\pi_{M^D} \circ S = \mathrm{id}_{M^D}$ and $(\pi_M)^D \circ S = \mathrm{id}_{M^D}$,
(ii) $S(\lambda \odot v) = \lambda \cdot (\lambda \odot S(v))$, for any $v \in M^D$, $\lambda \in R$.

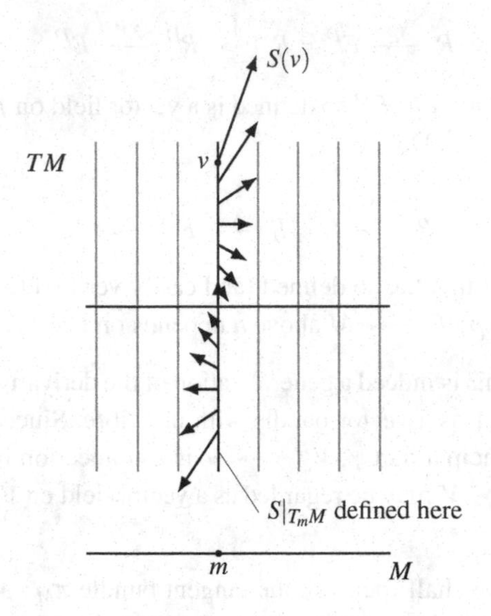

Figure 3.2 Spray

Definition 3.14 For a connection ∇ on the tangent bundle of M, a spray S on it is said to be a *geodesic spray* for ∇ if for every curve $a: R \longrightarrow M$, $a': R \longrightarrow M^D$ is an integral curve for S (i.e., a is a solution of the second order differential equation determined by S) if and only if a is a geodesic with respect to ∇.

Theorem 3.15 *Let M be infinitesimally linear.*

(i) *Given a connection* ∇ *on (the tangent bundle of)* M*, there is associated a geodesic spray* S_∇ *for* ∇*, given by the rule*

$$S_\nabla(v) = \nabla(v, v) .$$

(ii) *In addition, if* ∇ *and* ∇^* *are torsion free connections on* M *and* $S_\nabla = S_{\nabla^*}$*, then* $\nabla = \nabla^*$.

Proof (1). Given ∇, define S_∇ as the composite

$$M^D \xrightarrow{\text{diag}} M^D \times_M M^D \xrightarrow{\nabla} (M^D)^D .$$

We verify first that S_∇ is a spray:

(i) $$K \circ S_\nabla = K \circ \nabla \circ \langle \mathrm{id}_{M^D}, \mathrm{id}_{M^D} \rangle = \mathrm{id}_{M^D \times_M M^D} \circ \langle \mathrm{id}_{M^D}, \mathrm{id}_{M^D} \rangle = \langle \mathrm{id}_{M^D}, \mathrm{id}_{M^D} \rangle .$$

(ii) For $t \in M^D$, $\lambda \in R$,

$$S_\nabla(\lambda \odot t) = \nabla(\lambda \odot t, \lambda \odot t) = \lambda \cdot \nabla(\lambda \odot t, t) = \lambda \cdot (\lambda \odot \nabla(t,t)) = \lambda \cdot s(\lambda \odot S_\nabla(t)) .$$

Moreover, S_∇ is a geodesic spray for ∇. Indeed, a curve $a\colon R \longrightarrow M$ is a geodesic for ∇ if and only if a' is ∇-parallel, that is, if and only if $a'' = S_\nabla(a')$, and so, if and only if a' is an integral curve for S_∇.

(2). Let ∇ and ∇^* be torsion free connections such that $S_\nabla = S_{\nabla^*}$. We wish to show that $\nabla = \nabla^*$. By the equivalence between torsion free connections and torsion free connection maps, it is enough to show that $C = C^*$ for the associated connection maps. In turn, it is enough to prove that

$$H(t(0), C(t)) = H(t(0), C^*(t))$$

for every $t \in (M^D)^D$, since H is a monomorphism.

If $t \in (M^D)^D$ satisfies $(\pi_M)^D(t) = \pi_{M^D}(t)$, then the result is immediate as it is easy to verify. On this account, all we need to do is to reduce the general case to this special one, that is, to find, given an arbitrary $t \in (M^D)^D$, some $\tau \in (M^D)^D$ such that $(\pi_M)^D(\tau) = \pi_{M^D}(\tau)$ for which $C(t) = C(\tau)$ and $C^*(t) = C^*(\tau)$. Letting

$$\tau = t \odot \nabla(v, v \odot t(0))$$

where $v = (\pi_M)^D(t)$, it follows that $C(t) = C(\tau)$ from the linearity of C and the condition $C \cdot \nabla = 0$. Indeed,

$$C(t) = C(t \oplus \nabla(v, v \odot t(0))) \ominus C \circ \nabla(v, v \ominus t(0)) = C(\tau) .$$

In addition, τ satisfies the required conditions, for

$$(\pi_M)^D(t \oplus \nabla(v, v \ominus t(0))) = (\pi_M)^D(t) \oplus (\pi_M)^D(\nabla(v, v \ominus t(0))) = v \oplus 0 = v$$

and

$$\pi_{M^D}(t \oplus \nabla(v, v \ominus t(0))) = \pi_{M^D}(t \oplus \pi_{M^D}(\nabla(v, v \ominus t(0)))) = t(0) \ominus (v \ominus t(0)) = v .$$

It remains to be proved that $C^*(t) = C^*(\tau)$. Clearly this would be a consequence of $C^* \circ \nabla = 0$. We claim that the latter is the case. Recall that $S_\nabla = S_{\nabla^*}$. Using that the given connections and corresponding connection maps are torsion free, we establish the claim as follows. Notice that

$$C^* \circ \nabla\colon M^D \times_M M^D \longrightarrow M^D$$

is a bilinear form and that it vanishes on the diagonal since

$$\begin{aligned} C^*(\nabla(v,v)) &= C^*(S_\nabla(v)) = C^*(S_{\nabla^*}(v)) \\ &= C^*(\nabla^*(v,v)) = C^* \circ \nabla^*(v,v) = 0\,. \end{aligned}$$

To ease the notation, denote the composite $C^* \circ \nabla$ as φ.

We claim that

$$\varphi(v,w) = 0, \text{ for all } (v,w) \in M^D \times_M M^D\,.$$

Indeed,

$$\begin{aligned} 0 = \varphi(v+w,v+w) &= \varphi(v,v) + \varphi(v,w) + \varphi(w,v) + \varphi(w,w) \\ &= \varphi(v,w) + \varphi(w,v)\,. \end{aligned}$$

Therefore, $\varphi(v,w) = -\varphi(w,v)$ and so $2 \cdot \varphi(v,w) = 0$, from which it follows (since $2 \in Q$ is invertible and R is a Q-algebra) that $\varphi(v,w) = 0$ for all $(v,w) \in M^D \times_M M^D$. □

3.2 Local and Infinitesimal Exponential Map Property

In the classical proof of the Ambrose-Palais-Singer theorem [1, 98], the passage from a spray to a torsion free connection for which the spray is a geodesic spray is guaranteed by the local integrability of sprays, in turn a consequence of the corresponding theorem on the local existence of solutions to second-order differential equations [68]. For the purposes of explaining this remark, we state the following definition where the local structure for objects of a topos $\mathscr{E}$, where $(\mathscr{E},R)$ is a model of SDG, is assumed to be the intrinsic topological structure in the sense of Definition 1.20. In what follows also Exercise 1.21 shall be employed.

Definition 3.16 A *local flow* for a spray

$$S\colon M^D \longrightarrow (M^D)^D$$

on the tangent bundle of an infinitesimally linear object M of $\mathscr{E}$ is a pair (U,φ) with $U \in P(R \times M)$ such that

$$D \times M^D \subset U,$$

and $\varphi\colon U \longrightarrow M$ a morphism of $\mathscr{E}$ such that

(i) φ extends $\hat{S}\colon D \times M^D \longrightarrow M^D$, i.e., $\varphi(d,t) = S(t)(d)$.

(ii) For any (λ,t), (ξ,t), $(\lambda+\xi,t) \in U$,

$$(\lambda,\varphi(\xi,t)) \in U \text{ and } \varphi(\lambda+\xi,t) = \varphi(\lambda,\varphi(\xi,t)) .$$

(iii) For any $\lambda,\xi \in R$, $t \in M^D$,

$$\text{if } (\lambda\xi,t),(\lambda,\xi \odot t) \in U, \text{ then } \varphi(\lambda,\xi \odot t) = \xi.\varphi(\lambda\xi,t) .$$

Definition 3.17 An object M of $\mathscr{E}$ is said to have the *infinitesimal exponential map property* if given any spray

$$S\colon M^D \longrightarrow (M^D)^D$$

on M and a local flow (U,φ) for S in the sense of Definition 3.16, there is given a map

$$\exp_S\colon V \longrightarrow M,$$

with $V \in P(M^D)$ obtained via the pullback

$$\begin{array}{ccc} V & \longrightarrow & U \\ \downarrow & & \downarrow \\ M^D & \xrightarrow[\langle 1,\mathrm{id}\rangle]{} & R \times M^D \end{array},$$

such that

$$\exp_S(d \odot v) = v(d) .$$

Notice that since $(1,0_M) \in U$, $0_M \in V$ and so we can define $\exp_S\colon V \longrightarrow M$ by

$$\exp_S(t) = \pi_M(\varphi(1,t)) .$$

It follows easily from this that for any $\lambda \in R$ and $t \in M$ with $(\lambda,t) \in U$ and $\lambda \odot t \in V$, we have $\exp_S(\lambda \odot t) = \pi_M(\varphi(\lambda,t))$. In particular, for $d \in D$ and $t \in M$, $\exp_S(d \odot t) = t(d)$.

Notice that we have implicitly used that for $d \in D$ and $t \in M$, it follows that $d \odot t \in V$. The proof that it is so is an instructive exercise in the use of logical infinitesimals, so we include it below in more generality.

Lemma 3.18 *Let S be a spray on the tangent bundle of an infinitesimally linear object M of $\mathscr{E}$. Then, for any $n \geq 1$, given $(t_1,\dots,t_n) \in M^D \times_M \cdots \times_M M^D$ and $(d_1,\dots,d_n) \in V \times \cdots \times V$, it follows that $(d_1 \odot t_1) \oplus \cdots \oplus (d_n \odot t_n) \in V$.*

Proof Since $\neg\neg$ commutes with $\wedge$,

$$\neg\neg[(d_1,\ldots,d_n,t_1,\ldots t_n) = (0,\ldots,0,t_1,\ldots,t_n)]$$

and therefore

$$\neg\neg[(d_1 \odot t_1) \oplus \cdots \oplus (d_n \odot t_n) = 0_M] \ .$$

Since $0_M \in V$ and V is an intrinsic open of M^D, it follows that

$$(d_1 \odot t_1) \oplus \cdots \oplus (d_n \odot t_n) \in V \ .$$

□

With these, the formula

$$\nabla_S(v_1,v_2)(d_1,d_2) = \exp_S((d_1 \odot v_1) \oplus (d_2 \odot v_2))$$

can be shown to define a torsion free connection ∇_S such that the geodesic spray associated uniquely to ∇_S is S itself. The proof given in [28] is instructive for what follows in the next section, hence we include it below.

Proposition 3.19 *Let $(\mathscr{E},R)$ be a model of SDG, M an infinitesimally linear object of $\mathscr{E}$ and S a spray on the tangent bundle of M. Then there exists a torsion free connection ∇_S on the tangent bundle of M such that S is its geodesic spray.*

Proof For (U,φ) a local flow of S and $\exp_S$ the corresponding local exponential map, define

$$\nabla_S \colon M^D \times_M M^D \longrightarrow (M^D)^D$$

as the exponential adjoint of the composite

$$D \times M^D \times_M M^D \xrightarrow{\alpha} M^D \times_M M^D \xrightarrow{H} (M^D)^D \xrightarrow{\exp_S{}^D} M^D$$

where $\alpha(d,t_1,t_2) = (d \odot t_1, t_2)$ and $H(t,s)(d) = t \oplus (d \odot s)$. That ∇ is well defined follows from Lemma 3.18.

We now verify that the morphism ∇_S so defined is a connection.

(i) $K \circ \nabla_S = \text{id}$

follows from the identities

$$\begin{aligned}(\pi_{M^D} \circ \nabla_S)(t_1,t_2)(d) &= \nabla_S(t_1,t_2)(0)(d)\\ &= \exp_S((0 \odot t_1) \oplus (d \odot t_2))\\ &= \exp_S(d \odot t_2) = t_2(d)\\ &= \text{proj}_2(t_1,t_2)(d)\end{aligned}$$

and

$$\begin{aligned}((\pi_M)^D \circ \nabla_S)(t_1,t_2)(d) &= \pi_M(\nabla_S(t_1,t_2)(d)) \\ &= \nabla_S(t_1,t_2)(d)(0) \\ &= \exp_S((d \odot t_1) \oplus (0 \odot t_2)) \\ &= \exp_S(d \odot t_1) = t_1(d) \\ &= \mathrm{proj}_1(t_1,t_2)(d)\,.\end{aligned}$$

(ii) ∇_S is $\oplus$-linear .

To prove that

$$\nabla_S(t_1 \oplus t_2, s) = \nabla_S(t_1,s) \oplus \nabla_S(t_2,s)$$

we let $l\colon D(2) \longrightarrow M^D$ be given (well defined by Lemma 3.18) by

$$l(d_1,d_2) = \exp_S{}^D(H((d_1 \odot t_1) \oplus (d_2 \odot t_2), s))\,.$$

The result now follows from the easy verifications

$$l(d,d) = \nabla_S(t_1 \oplus t_2, s)(d),$$

$$l(d,0) = \nabla_S(t_1,s)(d),$$

and

$$l(0,d) = \nabla_S(t_2,s)(d)\,.$$

To prove that

$$\nabla_S(\lambda \odot (t,s)) = \lambda \odot \nabla_S(t,s)$$

we proceed as follows

$$\begin{aligned}\nabla_S(\lambda \odot (t,s))(d) &= \nabla_S(\lambda \odot t, s)(d) \\ &= \exp_S{}^D(H(d \odot (\lambda \odot t), s)) \\ &= \exp_S{}^D(d\lambda \odot t, s) \\ &= \nabla_S(t,s)(\lambda d) \\ &= (\lambda \odot \nabla_S(t,s))(d)\,.\end{aligned}$$

(iii) ∇_S is linear with respect to $+$.

To prove that

$$\nabla_S(t, s_1 + s_2)(d) = \nabla_S(t,s_1)(d) + \nabla_S(t,s_2)(d)$$

for $d \in D$, we let $r_d \colon D(2) \longrightarrow M$ be given (well defined by Lemma 3.18) by

$$r_d(d_1,d_2) = \exp_S((d \odot t) \oplus (d_1 \odot s_1) \oplus (d_2 \odot s_2))$$

and then verify that $r_d(d',d') = \nabla_S(t, s_1 \oplus s_2)(d)(d')$, whereas $r_d(d',0) = \nabla_S(t,s_1)(d)(d')$ and $r_d(0,d') = \nabla_S(t,s_2)(d)(d')$, for any $d' \in D$.

Next,

$$\lambda \odot \nabla_S(t,s)(d) = \lambda \odot {\exp_S}^D(H(d \odot t, s))$$

and

$$\begin{aligned}\nabla_S(t, \lambda \odot s)(d) &= {\exp_S}^D(H(d \odot t, \lambda \odot s)) \\ &= \lambda \odot {\exp_S}^D(H(d \odot t, s)),\end{aligned}$$

where the last identity uses that H and ${\exp_S}^D$ are both $\oplus$-linear.

(iv) S is a geodesic spray for ∇_S. Indeed,

$$\begin{aligned}\nabla_S(t,t)(d_1)(d_2) &= \exp_S((d_1 \odot t) \oplus (d_2 \odot t)) \\ &= \exp_S((d_1 + d_2) \odot t) = \pi_M \varphi_S(d_1 + d_2, t) \\ &= \pi_M \varphi_S(d_2, \varphi(d_1,t)) = \pi_M S(d_2, S(d_1,t)) \\ &= ((\pi_M)^D \circ S)(S(d_1,t))(d_2) \\ &= S(d_1,t)(d_2) = S(t)(d_1)(d_2)\end{aligned}$$

where we used that $(\pi_M)^D \circ S = \mathrm{id}_{M^D}$. Hence,

$$\nabla_S(t,t) = S(t)$$

as desired.

Note that for $(d_1,d_2) \in D(2)$, $\neg\neg[(d_1+d_2,t) = (0,t)]$ and $(0,t) \in U$, which is an intrinsic open, hence also $(d_1+d_2,t) \in U$ and so $\varphi(d_1+d_2,t)$ is meaningful.

(v) ∇_S is torsion free. Indeed,

$$\begin{aligned}\nabla_S(s,t)(d_1)(d_2) &= \exp_S(d_1 \odot s \oplus d_2 \odot t) \\ &= \exp_S(d_2 \odot t \oplus d_1 \odot s) \\ &= \nabla_S(t,s)(d_2)(d_1)\,.\end{aligned}$$

□

In this section we make the crucial observation that what is actually required of the local exponential map $\exp_S \colon V \longrightarrow M$ in the proof of Proposition 3.19

is its restriction to the subobject $D_2(M^D) \subset V$ given by the image of the morphism

$$\gamma\colon D \times D \times (M^D \times_M M^D) \longrightarrow M^D$$

whose rule is given by

$$(d_1, d_2, (v_1, v_2)) \mapsto (d_1 \odot v_1) \oplus (d_2 \odot v_2)\,.$$

An alternative (though less intuitive) way of proceeding is to let the infinitesimal exponential map associated to a spray S be defined directly on the domain of γ instead of on its image, but subject to suitable conditions, as in the following.

Definition 3.20 An object M of $\mathscr{E}$ is said to have *the infinitesimal exponential map property* if for any spray S on M there exists a morphism

$$e_S\colon D \times D \times (M^D \times_M M^D) \longrightarrow M$$

satisfying the following set of conditions for all $v \in M^D$, $d_1, d_2 \in D$, $\lambda \in R$ and $(v_1, v_2) \in M^D \times_M \times M^D$,

(EXP1)

$$\begin{aligned} e_S(d, 0, (v_1, v_2)) &= v_1(d) \\ e_S(0, d, (v_1, v_2)) &= v_2(d) \\ e_S(\lambda d_1, d_2, (v_1, v_2)) &= e_S(d_1, d_2, (\lambda \odot v_1, v_2)) \\ e_S(d_1, \lambda d_2, (v_1, v_2)) &= e_S(d_1, d_2, (v_1, \lambda \odot v_2) \end{aligned}$$

(EXP2)

$$e_S(d_2, d_1, (v_1, v_2)) = e_S(d_1, d_2, (v_2, v_1))$$

(EXP3)

$$e_S(d_1, d_2, (v, v)) = S(v)(d_1, d_2)\,.$$

The main result of this section is Theorem 3.22. It is the key to the second part of the proof of the Ambrose-Palais-Singer theorem within SDG as given in Theorem 3.23. We begin by stating a lemma whose proof depends on infinitesimal linearity and for which we refer to [28].

Lemma 3.21 *Let M be an infinitesimally linear object of $\mathscr{E}$, where $(\mathscr{E}, R)$ is a model of SDG. A spray on the tangent bundle of M is given equivalently by the data consisting of a 'spray map'*

$$\sigma\colon M^D \longrightarrow M^{D_2}$$

satisfying two conditions, as follows:
(i) $M^u \circ \sigma = \mathrm{id}$, where $u\colon D \hookrightarrow D_2$ is the inclusion, and
(ii) $\sigma(\lambda \odot v) = \lambda \odot \sigma(v)$, for any $v \in M^D$, $\lambda \in R$.

Theorem 3.22 *Let M be an infinitesimally linear object of $\mathscr{E}$. Then, M has the infinitesimal exponential map property.*

Proof The diagram of Weil algebras given by

$$Q[\varepsilon,\alpha] \xrightarrow{f} Q[\varepsilon,\alpha,\chi] \underset{h}{\overset{g}{\rightrightarrows}} Q[\varepsilon,\alpha,\chi,\mu]$$

where $\varepsilon^2 = \alpha^2 = \chi^3 = \mu^3 = 0$, $f(\varepsilon) = \varepsilon\cdot\chi$, $f(\alpha) = \alpha\cdot\chi$, $g(\varepsilon) = \varepsilon\cdot\mu$, $g(\alpha) = \alpha\cdot\mu$, $g(\chi) = \chi$, $h(\varepsilon) = \varepsilon\cdot\chi$, $h(\alpha) = \alpha\cdot\chi$ and $h(\chi) = \mu$ is an equalizer, as it is easily checked.

It follows that the diagram

$$(M^L)^{D\times D} \xrightarrow{(M^L)^F} (M^L)^{D\times D\times D_2} \underset{(M^L)^H}{\overset{(M^L)^G}{\rightrightarrows}} (M^L)^{D\times D\times D_2\times D}$$

with $L = M^D \times_M M^D$, which is infinitesimally linear since M is an equalizer in $\mathscr{E}$, where $F(d_1,d_2,\delta) = (\delta\cdot d_1, \delta\cdot d_2)$, $G(d_1,d_2,\delta_1,\delta_2) = (\delta_1\cdot d_1, \delta_1\cdot d_2, \delta)$ and $H(d_1,d_2,\delta_1,\delta_2) = (\delta_2\cdot d_1, \delta_2\cdot d_2, \delta_1)$, for all $d_1,d_2 \in D$ and $\delta_1,\delta_2 \in D_2$.

Let σ be the spray map associated with the spray S by Lemma 3.21 and define

$$e_\sigma : D\times D\times D_2\times L \longrightarrow M$$

by the identity

$$e_\sigma(d_1,d_2,\delta,(v_1,v_2)) = \sigma((d_1\odot v_1)\oplus(d_2\odot v_2))(\delta)\,.$$

It is easily verified that e_σ, regarded as a global section of $(M^L)^{D\times D\times D_2}$, equalizes the two morphisms in the above equalizer diagram, hence implies the existence of a unique global section of $(M^L)^{D\times D}$ which, when regarded as a morphism

$$e_S : D\times D\times L \longrightarrow M,$$

satisfies the condition

$$e_S(\delta\cdot d_1, \delta\cdot d_2,(v_1,v_2)) = e_\sigma(d_1,d_2,\delta,(v_1,v_2))\,.$$

The verifications of EXP1, EXP2 and EXP3 are routine.

□

We may now complete the proof of the Ambrose-Palais-Singer theorem within SDG whose first part was given in Theorem 3.15.

Theorem 3.23 *Let $(\mathscr{E},R)$ be a model of SDG with R a field of fractions. Let M be an infinitesimally linear object. Let S be a spray on the tangent bundle*

$\pi_M \colon M^D \longrightarrow M$ of M. Then there is defined a torsion free connection ∇_S on it such that S is geodesic spray associated (uniquely) to ∇_S.

Proof Since M is an infinitesimally linear object of $\mathscr{E}$, Theorem 3.22 guarantees the existence of a morphism

$$e_S \colon D \times D \times (M^D \times_M M^D) \longrightarrow M$$

satisfying conditions EXP1, EXP2 and EXP3 of Definition 3.20. For the given spray S, let

$$\nabla_S(v_1,v_2)(d_1,d_2) = e_S(d_1,d_2,(v_1,v_2)) .$$

(1) We verify first that ∇_S is a connection on the tangent bundle of M.

(a)
$$\nabla_S(v_1,v_2)(d,0) = e_S(d,0,(v_1,v_2)) = v_1(d)$$

and

$$\nabla_S(v_1,v_2)(0,d) = e_S(0,d,(v_1,v_2)) = v_2(d) .$$

(b)
$$\begin{aligned}(\lambda \cdot \nabla_S(v_1,v_2))(d_1,d_2) &= \nabla_S(v_1,v_2)(\lambda d_1,d_2)\\ &= e_S(\lambda d_1,d_2,(v_1,v_2))\\ &= e_S(d_1,d_2,(\lambda \odot v_1,v_2))\\ &= \nabla_S(\lambda \odot v_1,v_2)(d_1,d_2)\end{aligned}$$

and

$$\begin{aligned}(\lambda \odot \nabla_S(v_1,v_2))(d_1,d_2) &= \nabla_S(v_1,v_2)(d_1,\lambda d_2)\\ &= e_S(d_1,\lambda d_2,(v_1,v_2))\\ &= e_S(d_1,d_2,(v_1,\lambda \cdot v_2))\\ &= \nabla_S(v_1,\lambda \odot v_2)(d_1,d_2) .\end{aligned}$$

(2) We next verify that the connection ∇_S is torsion free.

$$\begin{aligned}\nabla_S(v_1,v_2)(d_2,d_1) = e_S(d_2,d_1,(v_1,v_2)) &= e_S(d_1,d_2,(v_2,v_1))\\ &= \nabla_S(v_2,v_1)(d_1,d_2) .\end{aligned}$$

(3) We end by proving that $S_{\nabla_S} = S$.

$$\begin{aligned}S_{\nabla_S}(v)(d_1)(d_2) = \nabla_S(v,v)(d_1,d_2) &= e_S(d_1,d_2,v,v)\\ &= S(v)(d_1,d_2) .\end{aligned}$$

Remark 3.24 That the classical Ambrose-Palais-Singer theorem [1] is recovered from Theorem 3.15 and Theorem 3.23 is a routine exercise given the existence of a well adapted model of SDG. A study of some of the models of SDG is given a systematic treatment in [89]. As an illustration of the power of SDG, explicit proofs of both the original Ambrose-Palais-Singer theorem and a generalization of it for function spaces are given in [89] as corollaries of the theorem as given within SDG.

A theory of connections within SDG may be pursued further. An excellent classical reference is [98]. In Chapter 3 of [98], it is shown that a Riemannian metric on a manifold determines a natural affine connection, the Levi-Civita connection. This operator is of fundamental importance in the study of the geometry of the metric. In Chapter 5 of [98] about the isometry group of the manifold, it is shown how the study of the Killing fields on a Riemannian manifold tells us something about the isometry group of the manifold. There are other 'geometric' vector fields to consider on manifolds. Some of these venues may require additional axioms to those of SDG and, as usual, such are permitted insofar as they are shown to be valid in some well adapted model.

4
Calculus of Variations in SDG

In this chapter, based on [52, 27], we shall give examples drawn from the classical calculus of variations which are conceptually advantageous to deal with by embedding the category of smooth manifolds into a well adapted model of SDG. The main differences between SDG and classical differential geometry in general are due to the availability of both infinitesimals and of arbitrary function spaces in the former but not in the latter. In order to carry out our program for the calculus of variations, some special axioms are added to SDG. As in the previous chapter on connections and sprays, the existence of a well adapted model of SDG, which in this case should also satisfy the additional axioms employed, is what connects it with the classical calculus of variations. That such a model exists is shown in the last part of this book. The main feature in the calculus of variations within SDG that makes it into a more natural theory than the classical one is that in the former the theory of extrema of functionals is a particular case of a general theory of critical points and that, just as in the theory of connections in the previous chapter, we work here with infinitesimally linear objects in a topos, a class closed under fibred products, exponentiation and étale descent. In particular, the range of applicability of the calculus of variations within SDG extends well beyond the classical one.

4.1 Basic Questions of the Calculus of Variations

Among the most typical questions of the variational calculus [49, 87] are the following two: find the shortest curve between two points on a surface, and find a closed curve of a given length and maximal enclosed area.

If M is a (finite dimensional) smooth manifold and $f : M \longrightarrow \mathbb{R}$ is a smooth function, maxima and minima are found among the critical points of f, that is, among those points $x \in M$ for which $f'(x) = 0$. This condition often brings about useful characterizations of the extremal points. For functionals of the form $F\colon M^{[0,1]} \longrightarrow \mathbb{R}$, such as length, or area, the 'derivative' of F can be given a meaning but in general need not exist. When it does, it can be used for the same purpose, namely to detect the 'critical paths' of functionals. However, these are *ad hoc* notions which require the introduction of a new concept, to wit, that of a 'variation'.

To be more precise, if $\Lambda_{p,q}$ denotes the set of paths $c\colon [0,1] \longrightarrow \mathbb{R}^n$ from p to q, for $p,q \in \mathbb{R}^n$, that is, of those smooth c such that $c(0) = p$ and $c(1) = q$, then, for any functional $E\colon \Lambda_{p,q} \longrightarrow \mathbb{R}$, for instance, the 'action integral'

$$E(c) = \int_0^1 ||c'(t)||^2 dt\,,$$

the 'principle of least action' states that E will be minimized among those paths c traversing a geodesic, that is, among those paths c for which $c''|_{[0,1]} = 0$. Here is where 'variations' come in.

A *variation* of $c \in \Lambda_{p,q}$, keeping the endpoints fixed, is a map

$$\alpha\colon (-\varepsilon,\varepsilon) \longrightarrow \mathbb{R}^{n[0,1]},$$

with $\varepsilon > 0$, such that,

(i) $\overline{\alpha}\colon (-\varepsilon,\varepsilon) \times [0,1] \longrightarrow \mathbb{R}^n$ is smooth,
(ii) $\forall u \in (-\varepsilon,\varepsilon)\ (\overline{\alpha}(u,0) = p \wedge \overline{\alpha}(u,1) = q)$, and
(iii) $\forall t \in [0,1]\ (\overline{\alpha}(0,t) = c(t))$.

Using this notion one defines

$$dE|_c\colon T_c(\Lambda_{p,q}) \longrightarrow \mathbb{R}$$

where $T_c(\Lambda_{p,q})$ is the *tangent space* of $\Lambda_{p,q}$ at c, that is, the space of vector fields w over c with $w(0) = w(1) = 0$, as follows. First, one uses the classical theory of differential equations to find a variation

$$\overline{\alpha}\colon (\varepsilon,\varepsilon) \times [0,1] \longrightarrow \mathbb{R}^n$$

which is a local solution of the differential equation

$$\frac{\partial}{\partial u}\overline{\alpha}(u,t)|_{u=0} = w(t)$$

subject to the conditions $\overline{\alpha}(u,0) = p$ and $\overline{\alpha}(u,1) = q$, as well as $\overline{\alpha}(0,t) = c(t)$. This is well defined provided that the derivative in question exists and is independent of the choices of α.

A *critical path* for E then is a $c \in \Lambda_{p,q}$ such that

$$\frac{d}{du}E(\alpha(u))|_{u=0} = 0.$$

For instance, the critical paths for the Euler operator are the solutions to the Euler-Lagrange equations. Consider a dynamical system as given by a Lagrangian $\mathscr{L}$ in the study of a continuous body **B** during a time lapse $[a,b]$. Associated with $\mathscr{L}$ is a functional

$$\overline{\mathscr{L}}_a^b \colon \mathbf{Q}^{[a,b]} \longrightarrow \mathbb{R}$$

where $\mathbf{Q}$ is the configuration space of **B** and $\mathbf{Q}^{[a,b]}$ is the space of smooth paths in the configuration space. The formula defining $\overline{\mathscr{L}}_a^b$ is

$$\overline{\mathscr{L}}_a^b(q) = \int_a^b \mathscr{L}(q,\dot{q})dt$$

where $q \in \mathbf{Q}^{[a,b]}$ and $\dot{q}$ is the velocity of the curve q. In general, $\mathscr{L}$ is defined on the state space **X** of **B** whose elements are pairs (q,v) consisting of a configuration q and a velocity vector v, with $\mathscr{L} \colon \mathbf{X} \longrightarrow \mathbb{R}$ interpreted as 'the work needed to be added to the potential energy of q in order to achieve the kinetic energy v'. Possible motions of **B** in the time lapse $[a,b]$ are among those paths q for which $\overline{\mathscr{L}}_a^b(q)$ is minimal, and the corresponding classical result says that this is the case precisely when q is a solution of the Euler-Lagrange equations

$$\left.\frac{\partial f}{\partial q} - \frac{d}{dt}\frac{\partial f}{\partial \dot{q}}\right|_{[a,b]} = 0.$$

We next highlight, before proceeding, certain points that will be illustrated in our treatment within SDG of classical differential geometry in general and of the calculus of variations in particular.

Remark 4.1 (i) In the context of SDG, the notions of tangent bundle of a space of paths and that of the derivative of a functional are not *ad hoc*—they always exist. In particular, the notion of a critical path is meaningful.

(ii) In the context of SDG, no limits are involved in the actual computation of the value of a derivative of a function or of a functional. This procedure is, thanks to the existence of non-trivial nilpotent elements in the line, entirely algebraic. In particular, in the context of SDG, variations are not needed except in their infinitesimal guise. Thus, one needs not integrate a vector field or solve a differential equation locally, as one may work with infinitesimal notions directly. Moreover, this can be done for a wide class of objects of $\mathscr{E}$—to wit, the infinitesimally linear objects (or 'generalized manifolds').

(iii) In the presence of a well adapted model $(\mathscr{S},R)$ of SDG, the class of infinitesimally linear objects in $\mathscr{E}$ is closed under inverse limits and exponentiation by arbitrary objects and includes not just all smooth manifolds but also spaces with singularities and spaces of smooth functions. In other words, an infinitesimally linear object M of $\mathscr{E}$ behaves, at least with respect to maps from infinitesimal spaces into it, as if it had local coordinates.

We next examine, within SDG, the usual identification of the critical paths for the energy integral (associated with a metric) with the geodesics.

Let $(\mathscr{E}, R)$ be a model of SDG. In particular, $\mathscr{E}$ is a topos, R is a commutative ring with 1 in it, and Axiom J is internally valid in $\mathscr{E}$. We will be using here mostly the 'Kock-Lawvere axiom', a particular case of Axiom J which we recall next.

Axiom 4.2 (KL-axiom) *The morphism*

$$\alpha_R \colon R \times R \longrightarrow R^D$$

defined as $(a,b) \mapsto [d \mapsto a + d \cdot b]$ *for* $a, b \in R$, $d \in D$, *is an isomorphism.*

For a morphism $f \colon R \longrightarrow R$, its derivative $f'(x)$ at some $x \in R$ is defined, on account of the KL-axiom, by means of the formula

$$\forall d \in D\, [f(x+d) = f(x) + d \cdot f'(x)] \, .$$

Recall from Chapter 2 (or see [61]) that not only do all usual rules for derivatives follow from it, but also that this can be generalized in various ways to include partial derivatives of functions defined on suitable subobjects of R^n with values on an R-module V that is required to satisfy 'the vector form of the KL-axiom', which says that the map $\alpha_V \colon V \times V \longrightarrow V^D$, defined as above for $a, b \in V$, is an isomorphism.

If M is an infinitesimally linear object of $\mathscr{E}$, then M^D, together with the 'evaluation at 0' map $\pi_M \colon M^D \longrightarrow M$, can be thought of as the tangent bundle of M. Fibrewise, M^D is an R-module. If V is also infinitesimally linear and an R-module which satisfies the vector form of the KL-axiom, then for any $y \in V$,

$$T_y V =_{\text{def}} {\pi_1}^{-1}\{y\} \cong V \, .$$

For M an infinitesimally linear object in $\mathscr{E}$ and $f \colon M \longrightarrow V$ any morphism of $\mathscr{E}$, the corresponding morphism

$$f^D \colon M^D \longrightarrow V^D$$

of $\mathscr{E}$ restricts, for any $x \in M$, to an R-linear map

$$(df)_x \colon T_x M \longrightarrow T_{f(x)} V \cong V$$

whose defining equation is given by the formula, for $v \in T_x M$,

$$\forall d \in D [f(v(d)) = f(x) + d \cdot (df)_x(v)] \, .$$

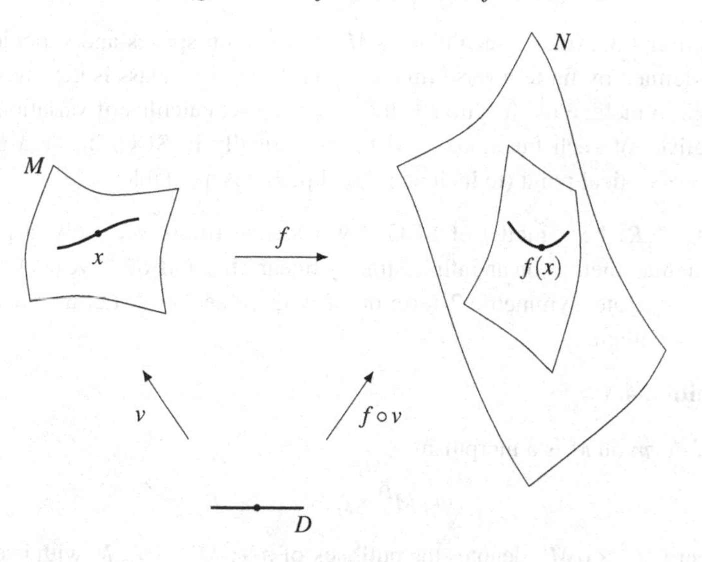

Figure 4.1 Exponentiating to D

For instance, if $f\colon R \longrightarrow R$, using the KL-axiom and that 'universally quantified $d \in D$ can be cancelled', a principle that emanates from it, one obtains the following identity

$$f'(x) = (df)_x(v_x),$$

where $v_x \in T_xR$ is the canonical tangent vector at x given by $d \mapsto x+d$.

Next, one can obtain the identity

$$(df)_x(v) = v'(0) \cdot f'(x)$$

for any $v \in T_xR$ by two applications of the KL-axiom

$$f(v(d)) = f(v(0)+d \cdot v'(0)) = f(x+d \cdot v'(0)) = f(x)+d \cdot v'(0) \cdot f'(x)$$

using, for the last identity, that $d \cdot v'(0) \in D$. In particular, from the last two identities we obtain that, for any $x \in M$, $f'(x) = 0$ if and only if $(df)_x = 0$. These considerations suggest the following definition.

Definition 4.3 Let $(\mathscr{E}, R)$ be a model of SDG. Let M and V be infinitesimally linear objects of $\mathscr{E}$, with V an R-module that satisfies the vector form of the KL axiom. Say that $x \in M$ is *a critical point* of $f\colon M \longrightarrow V$ if $(df)_x = 0$.

Remark 4.4 The notion of derivative in SDG is uniformly given for any map $f\colon M \longrightarrow V$, where M is just required to be infinitesimally linear and V as in

Definition 4.3. Among such objects M are function spaces and subobjects of them defined by finite inverse limit constructions. The class is therefore large enough to include the functionals that occur in the calculus of variations. The derivatives of such functionals exist automatically in SDG, hence a general notion of critical point (to include critical paths) is available.

Let $(\mathscr{E}, R)$ be a model of SDG. By a *metric* (more precisely, a pseudo-Riemannian metric) on an infinitesimally linear object M of $\mathscr{E}$ we shall mean a non-degenerate, symmetric 2-form on M with values on R. Let us make these notions explicit.

Definition 4.5

- A *2-form* on M is a morphism

$$\omega\colon M^D \times_M M^D \longrightarrow R$$

  where $M^D \times_M M^D$ denotes the pullback of $\pi_M\colon M^D \longrightarrow M$ with itself.
- A 2-form ω on M is said to be *symmetric* if $R^{\bar{\sigma}}(\omega) = \omega$ for any $\sigma \in S_2$, the symmetric group in 2 letters, with $\bar{\sigma}\colon M^D \times_M M^D \longrightarrow M^D \times_M M^D$ defined by $\bar{\sigma}\pi_i = \pi_{\sigma(i)}$ for $i = 1, 2$.
- A 2-form ω on M is said to be *non-degenerate* if the morphisms ω_1 and ω_2 obtained from ω by fixing the first, respectively the second, variable, are both isomorphisms.

For M any infinitesimally linear object, and $p, q \in M$, denote by

$$\Lambda_{p,q} = [[c \in M^{[0,1]} \mid c(0) = p \wedge c(1) = q]]$$

the object of paths in M beginning at p and ending at q. Since M is infinitesimally linear, so is $\Lambda_{p,q}$. Furthermore, the tangent bundle

$$\pi_{\Lambda_{p,q}}\colon {\Lambda_{p,q}}^D \longrightarrow \Lambda_{p,q}$$

is fibrewise R-linear. For $c \in \Lambda_{p,q}$ defined at stage X, let $c_* \in \Lambda_{p,q}$ be the corresponding velocity field , that is, the composite

$$X \xrightarrow{c} M^{[0,1]} \xrightarrow{(-)^D} (M^D)^{([0,1]^D)} \xrightarrow{M^{Dj}} (M^D)^{[0,1]}$$

where $j\colon [0,1] \longrightarrow [0,1]^D$ is the transpose of the addition map, that is, $j(t)(d) = j(t+d)$.

Consider the particular case where $M = R^n$ and ω is the *canonical metric* on it defined as follows :

$$\omega\colon R^{nD} \times_{R^n} R^{nD} \cong R^{2n} \times_{R^n} R^{2n} \longrightarrow R$$

defined, for $(a_1,\ldots,a_{2n}) \in R^{2n}$ and $(b_1,\ldots,b_{2n}) \in R^{2n}$, with $a_i = b_i$ for $i = 1,2,\ldots,n$, by

$$\omega(a_1,\ldots,a_{2n},b_1,\ldots,b_{2n}) = \sum_{i=n+1}^{2n} a_i b_i.$$

In particular, for $M = R^n$ and ω the canonical metric on it, $\omega(c_*,c_*)$ is meaningful and an element of $R^{[0,1]}$. We may express

$$c_* = (c_1,\ldots,c_n,c'_1,\ldots,c'_n),$$

using the Kock-Lawvere axiom. Using it, we have the formula

$$\omega(c_*,c_*) = \sum_{i=1}^{n} (c'_i)^2.$$

For the purposes of a calculus of variations within SDG, we now introduce some special axioms. Let $(\mathscr{E},R)$ be a model of SDG—that is, Axiom J, Axiom W, Postulate K, Postulate F and Postulate O all hold in $\mathscr{E}$.

Axiom 4.6 (Axiom I)

$$\forall f \in R^{[0,1]}\ \exists! g \in R^{[0,1]}\ \left[g' = f \wedge g(0) = 0 \right].$$

(Denoting $g(t) = \int_0^t f(u)du$, it can be shown that all the usual properties follow from the axiom of integration [61].)

Axiom 4.7 (Axiom C)

$$\forall f \in R^{[0,1]}\ \forall t \in (0,1) f(t) > 0 \Rightarrow \exists a,b \in R\ \big[0 < a < t < b < 1 \\ \wedge\ \forall u \in (a,b)\,(f(u) > 0) \big].$$

Axiom 4.8 (Axiom X)

$$\forall a < b \in R\ \exists h \in R^R\ \forall t \in R\ \big[(a < t < b) \Rightarrow h(t) > 0 \\ \wedge\ (t < a \vee t > b) \Rightarrow h(t) = 0 \big].$$

Axiom 4.9 (Axiom P)

$$\forall f \in R^{[0,1]}\ \forall t \in (0,1)\ \left[f(t) > 0 \Rightarrow \int_0^1 f(t)dt > 0 \right].$$

In the context of SDG, the *energy* of a path $c \in \Lambda_{p,q}$ is now defined as usual,

$$E(c) = \int_0^1 \omega(c_*,c_*)dt$$

for any metric ω.

Remark 4.10 If ω is the canonical metric on R^n, then we have

$$E(c) = \int_0^1 ||c''(t)||^2 dt.$$

As a morphism in the topos $\mathscr{E}$,

$$E: \Lambda_{p,q} \longrightarrow R$$

and since $\Lambda_{p,q}$ is infinitesimally linear,

$$(dE)_c: T_c(\Lambda_{p,q}) \longrightarrow R^D$$

exists as an R-linear map and is given simply as the restriction of

$$E^D: {\Lambda_{p,q}}^D \longrightarrow R^D$$

to the fibre above c.

An explicit description of the fibre $T_c(\Lambda_{p,q})$ above c of the tangent bundle of $\Lambda_{p,q}$ is given by the extension of the formula

$$[[v \in R^{nD\times[0,1]} \mid \forall t \in [0,1] \, \big(v(0,t) = c(t)\big) \\ \wedge \, \forall d \in D \, \big(v(d,0) = p \wedge v(d,1) = q\big)]],$$

where we have used the same notation v whether it is an element of $(R^{n[0,1]})^D$ or of its equivalent object $R^{nD\times[0,1]}$.

Recall that c is a critical point of E if $(dE)_c = 0$. Call c a *geodesic* if

$$\bigwedge_{i=1}^{n} (\forall t \in [0,1] \;\; c_i''(t) = 0).$$

Using the internal logic of the topos $\mathscr{E}$, one can define corresponding subobjects $\mathrm{Crit}(\Lambda_{p,q})$ and $\mathrm{Geod}(\Lambda_{p,q})$ as subobjects of $\Lambda_{p,q}$. The classical result states that, for global sections,

$$\mathrm{Crit}(\Lambda_{p,q}) = \mathrm{Geod}(\Lambda_{p,q}).$$

In the SDG context we shall 'almost' derive the same result but without any restrictions on the elements of $\Lambda_{p,q}$. Before discussing this question, we shall explicitly give the synthetic reasonings.

Lemma 4.11 *(Formula for the first variation of arc length) The following holds in* $\mathscr{E}$

$$\forall p,q \in R^n \; \forall c \in \Lambda_{p,q} \;\; \forall v \in T_c(\Lambda_{p,q}) \\ (dE)_c(v) = -2\int_0^1 \sum_{i=1}^{n} \big(v_i'(0)(t) \cdot c_i''(t)\big) dt.$$

Proof By definition, for $d \in D$,

$$E(v(d)) = \int_0^1 \sum_{i=1}^n \big(\frac{\partial}{\partial t} v_i(d,t)\big)^2 dt.$$

By the Kock-Lawvere axiom applied to $\frac{\partial}{\partial t} v_i(-,t) \in R^D$, the above equals

$$\int_0^1 \sum_{i=1}^n \big(\frac{\partial}{\partial t} v_i(0,t) + d \cdot \frac{\partial}{\partial d}\frac{\partial}{\partial t} v_i(d,t)|_{d=0}\big)^2 dt.$$

Using now that $d^2 = 0$ in the binomial expansion of the binary sum, and interchanging the order of the partial derivatives, the above is, in turn, equal to

$$\int_0^1 \sum_{i=1}^n (c_i'(t))^2 dt + d \cdot 2 \int_0^1 \sum_{i=1}^n \big(c_i'(t) \cdot \frac{d}{dt} v_i'(0)(t)\big) dt.$$

It follows that

$$(dE)_c(v) = 2 \int_0^1 \sum_{i=1}^n \big(c_i'(t) \cdot \frac{d}{dt} v_i'(0)(t)\big) dt.$$

Integrating by parts, this becomes

$$(dE)_c(v) = 2 \sum_{i=1}^n c_i'(t) \cdot v_i'(0)(t)|_{t=0}^{t=1} - 2 \int_0^1 \sum_{i=1}^n \big(v_i'(0)(t) \cdot c_i''(t)\big) dt.$$

Notice that for any $v \in T_c(\Lambda_{p,q})$,

$$v_i'(0) = 0 = v_i'(1) = 0$$

for every $i = 1, \ldots, n$ and $v = (v_1, \ldots, v_n)$ with $v_i \in (R^D)^{[0,1]}$. It follows then that

$$(dE)_c(v) = -2 \int_0^1 \sum_{i=1}^n \big(v_i'(0)(t) \cdot c_i''(t)\big) dt.$$

□

Corollary 4.12 *The following holds in $\mathscr{E}$:*

$$\forall p, q \subset R^n \; \forall c \in \Lambda_{p,q} \; \big(c \in \mathrm{Geod}(\Lambda_{p,q}) \Rightarrow c \in \mathrm{Crit}(\Lambda_{p,q})\big).$$

Our next task will be to establish the fundamental lemma of the calculus of variations, to be used in establishing an almost converse to Lemma 4.12.

Lemma 4.13 *(Fundamental Lemma of the Calculus of Variations) The following holds in $\mathscr{E}$ for all $f \in R^{[0,1]}$:*

$$(\exists t \in [0,1]\, \neg(f(t)=0)) \Rightarrow \exists h \in R^R \Big[h(0)=0 \wedge h(1)=0 \\ \wedge \neg\Big(\int_0^1 h(x)f(x)dx = 0\Big)\Big].$$

Proof Let $f \in R^{[0,1]}$. Assume

$$\exists t \in [0,1]\, \neg(f(t)=0).$$

Using Postulates K and O, the assumption is equivalently stated as

$$\exists t \in [0,1]\, (f(t) > 0 \vee -f(t) > 0).$$

From

$$\exists t \in [0,1] \ (f(t) > 0)$$

and Postulate C follows that

$$\exists a,b \in R \ [0 < a < b < 1 \wedge \forall t \in (a,b) f(t) > 0].$$

In turn, from it and the Postulate X follows

$$\exists a,b \in R\, \exists h \in R^R \ [\ (0 < a < b < 1) \wedge \forall t \in (a,b)\ (h(t)\cdot f(t) > 0) \\ \wedge \forall t \in R\, ((t < a \vee t > b) \Rightarrow h(t) = 0)\ ].$$

In turn, using the above and Postulate C, we get

$$\exists a,b \in R\, \exists h \in R^R \Big[\ (0 < a < b < 1) \wedge \int_a^b h(t)f(t)dt > 0 \\ \wedge \forall t \in R\, ((t < a \ \vee\ t > b) \Rightarrow h(t) = 0)\Big].$$

Finally, properties of the integral yield

$$\exists h \in R^R \Big(\int_0^1 h(t)f(t) > 0 \wedge h(0) = 0 \wedge h(1) = 1 \Big).$$

A similar argument but starting instead with

$$\exists t \in [0,1]\ (-f(t) > 0)\ ,$$

and replacing h by $-h$ above, yields the same result. It is now a deduction rule of Heyting logic which allows us to conclude the desired result. □

Definition 4.14 For $p,q \in R$ a curve $c \in \Lambda_{p,q}$ is said to be *almost a geodesic* if the following holds:

$$\bigwedge_{i=1}^{n} \forall t \in [0,1] \neg\neg (c_i''(t) = 0).$$

Denote by $\mathrm{Geod}_{\neg\neg}(\Lambda_{p,q})$ the subobject of $\Lambda_{p,q}$ which is the extension of the above formula in $\mathscr{E}$.

Remark 4.15 Recall the terminology $x\#0$, short for the statement 'x is invertible', where $x \in R$. It follows from the field axiom and the rules of Heyting logic that an equivalent notion to the one introduced in Definition 4.14 is the following

$$c \in \mathrm{Geod}_{\neg\neg}(\Lambda_{p,q}) \leftrightarrow \bigwedge_{i=1}^{n} \neg\exists t \in [0,1]\, c_i''(t)\#0.$$

Theorem 4.16 *The following statement holds in* $\mathscr{E}$*:*

$$\forall p,q \in R^n\, \forall c \in \Lambda_{p,q}\, [c \in \mathrm{Crit}(\Lambda_{p,q}) \Rightarrow c \in \mathrm{Geod}\neg\neg(\Lambda_{p,q})].$$

Proof Let $c \in \mathrm{Crit}(E)$. Assume that for a given $i \in \{1,\dots,n\}$, we have

$$\exists t \in [0,1]\ (c_i''(t)\#0).$$

From it follows that

$$\exists t \in [0,1]\ (c_i''(t)^2\#0).$$

We have

$$\exists h \in (R^R)\ [h(0) = 0 \wedge h(1) = 1 \wedge \int_0^1 h(t)\cdot c_i''(t)^2 dt\, \#\, 0].$$

Define $v \in (R^n)^{D\times[0,1]}$ by

$$v(d,t) = (c_1(t),\dots,c_i(t) + d\cdot h(t)\cdot c_i''(t),\dots,c_n(t))$$

for $d \in D$, $t \in [0,1]$. It is easy to check that

$$v \in T_c(\Lambda_{p,q}).$$

Furthermore it is the case that $v'(0)(t) = 0$ if $j \neq i$ and $v'(0)(t) = h(t)\cdot c_i''(t)$ if $j = i$. It now follows from Lemma 4.11 and the above, that

$$\begin{aligned}(dE)_c(v) &= -2\int_0^1 \sum_{i=1}^{n} (v_i''(0)(t)\cdot c_i''(t))dt \\ &= -2\int_0^1 h(t)\cdot c_i''(t)^2 dt\#0.\end{aligned}$$

This contradicts the assumption that $(dE)_c = 0$. Therefore

$$\neg \exists t \in [0,1]\ (c_i''(t) \# 0)\,.$$

This concludes the proof. □

Remark 4.17 As we have already observed, in SDG there is no need to resort to variations in order to deal with critical paths and geodesics. On the other hand it would *a priori* seem that the result obtained is not quite what one wants in the classical case. Indeed, the critical paths are almost the geodesics. In other words, although every geodesic is necessarily a critical path, it is not the case that every critical path is a geodesic but almost a geodesic. This means that to assert that every critical path is not a geodesic yields a contradiction. We now argue as follows. Let $(\mathscr{E}, R)$ be a well adapted model of all the axioms assumed in this chapter, together with the inclusion

$$i\colon \mathscr{M}^\infty \hookrightarrow \mathscr{E}$$

from the category of smooth manifolds into $\mathscr{E}$, preserving transversal pullbacks and coverings, and sending $\mathbb{R}$ into R. We claim that for *actual* paths $c \in \Lambda_{p,q}$, $p, q \in \mathbb{R}^n$, regarded in $\mathscr{E}$ via the above inclusion, it is actually the case that c is almost a geodesic if and only if c is a geodesic. Indeed, the condition stating that c is almost a geodesic translates, for each $i = 1, \ldots, n$ into a factorization of $c_i''\colon [0,1] \longrightarrow R$ through the subobject $\Delta \hookrightarrow R$, where $\Delta = \neg\neg\{0\}$. However, the only global section of Δ is $0\colon 1 \longrightarrow \Delta$ and therefore, since actual paths become global sections in the passage from $\mathscr{M}^\infty$ to $\mathscr{E}$, c_i'' must be the zero morphism $[0,1] \longrightarrow R$ for each i. In conclusion, when Corollary 4.12 and Theorem 4.16 of the calculus of variations within SDG are applied to *classically constructed objects*, they do give the well known classical results. We conclude that (at least in this case) constructive mathematics enriches (but does not detract from) classical mathematics.

4.2 The Euler-Lagrange Equations

As in [72], we shall use certain assumptions of a categorical nature in the study of the dynamics of a 'continuous body' **B** as given by a Lagrangian

$$\mathscr{L}\colon \mathbf{X} \longrightarrow \mathbb{R}$$

defined on the 'state space' **X** of **B**.

Specifically, working now inside a model $(\mathscr{E}, R)$ of Axiom J (Axiom 4.2), we assume that $X = Q^D$ where Q is the 'configuration space' of B. Furthermore we shall assume that Q is an infinitesimally linear R-module satisfying the vector form of the KL-axiom. Associated to the Lagrangian and for each 'time lapse' $[a,b]$ is a morphism

$$\overline{\mathscr{L}}_a^b \colon Q^{[a,b]} \longrightarrow R\,,$$

defined as the composite

$$Q^{[a,b]} \xrightarrow{\text{can}^{[a,b]}} (Q^D)^{[a,b]} \xrightarrow{\mathscr{L}^{[a,b]}} R^{[a,b]} \xrightarrow{\int_a^b} R\,,$$

where can(q) is given by the law $t \mapsto [d \mapsto q(t+d)]$, for $q \in Q^{[a,b]}$, $t \in [a,b]$, $d \in D$.

Using the vector form of the KL-axiom,

$$Q^D \cong Q \times Q,$$

so that a coordinate expression for an element of $Q^{[a,b]}$ is a pair $(q,\dot{q})$ where, for $q \in Q^{[a,b]}$, $t \in [a,b]$ and $d \in D$, $q(t+d) = q(t) + d\dot{q}(t)$. With this notation we write, in more familiar form

$$\overline{\mathscr{L}}_a^b(q) = \int_a^b \mathscr{L}(q(t),\dot{q}(t))dt\,.$$

On account of the usual interpretation of the Lagrangian and of its corresponding action integral (see § 4.1), the object of 'possible motions of B in the time lapse $[a,b]$' can in our context be taken to be the subobject of $Q^{[a,b]}$ consisting of those paths q which are critical for $\overline{\mathscr{L}}_a^b$, that is, the object

$$\text{Crit}(\overline{\mathscr{L}}_a^b) = [[q \in Q^{[a,b]} \mid (d\overline{\mathscr{L}}_a^b)_q = 0]]\,.$$

Lemma 4.18 *Let $(\mathscr{E},R)$ be a model of SDG which in addition satisfies Axioms I, C, X and P. Then the following holds in $\mathscr{E}$:*

$$\begin{aligned}\forall v \in (Q^{[a,b]})^D \Big[\overline{\mathscr{L}}_a^b(v) = \int_a^b \Big(&\frac{\partial}{\partial q}\mathscr{L}(v(0,t),\frac{\partial}{\partial t}v(0,t)) \\ &- \frac{d}{dt}\frac{\partial}{\partial \dot{q}}\mathscr{L}(v(0,t))\Big)\cdot v'(0)(t)dt \\ &+ \frac{\partial}{\partial \dot{q}}\mathscr{L}(v(0,t),\frac{\partial}{\partial t}v(0,t))\cdot v'(0)(t)\big|_{t=a}^{t=b}\Big]\,.\end{aligned}$$

Proof Explicitly, we have

$$\begin{aligned}\overline{\mathscr{L}_a^b}(v(d)) &= \int_a^b [t \mapsto \mathscr{L}[\delta \mapsto v(d)_\star(t)(\delta)]]dt \\ &= \int_a^b [t \mapsto \mathscr{L}[\delta \mapsto v(0,t+d) + d \cdot \frac{\partial}{\partial d} v(d,t+\delta)|_{d=0}]]dt \\ &= \int_a^b [t \mapsto \mathscr{L}[\delta \mapsto v(0,t) + \delta \cdot \frac{\partial}{\partial t} v(0,t) + d \cdot \frac{\partial}{\partial d} v(d,t)|_{d=0} \\ &\qquad + d \cdot \delta \cdot \frac{\partial}{\partial t}\frac{\partial}{\partial d} v(d,t)|_{d=0}]]dt.\end{aligned}$$

where $d, \delta \in D$ and $t \in [a,b]$. Using the coordinate expression for $\mathscr{L}$, the above can be expressed as

$$\begin{aligned}&\int_a^b [t \mapsto \mathscr{L}(v(0,t) + d \cdot \frac{\partial}{\partial d} v(d,t)|_{d=0}, \frac{\partial}{\partial t} v(0,t) + d \cdot \frac{\partial}{\partial t}\frac{\partial}{\partial d} v(0,d)|_{d=0})]dt \\ &= \int_a^b [t \mapsto \mathscr{L}(v(0,t), \frac{\partial}{\partial t} v(0,t) + d \cdot \frac{\partial}{\partial t}\frac{\partial}{\partial d} v(d,t)|_{d=0} \cdot \frac{\partial}{\partial d} v(0,t)|_{d=0} \\ &\qquad \cdot \frac{\partial \mathscr{L}}{\partial q}(v(0,t), \frac{\partial}{\partial t} v(0,t)) + d \cdot \frac{\partial}{\partial t}\frac{\partial}{\partial d} v(d,t)|_{d=0})]dt\end{aligned}$$

by the KL-axiom and, employing it again twice and using that $d^2 = 0$ we get

$$\begin{aligned}&\int_a^b \mathscr{L}(v(0,t), \frac{\partial}{\partial t} v(0,t))dt + d \cdot \int_a^b \Big(\frac{\partial \mathscr{L}}{\partial q}(v(0,t), \frac{\partial}{\partial t} v(0,t)) \cdot v'(0)(t) \\ &\qquad + \frac{\partial \mathscr{L}}{\partial \dot{q}}(v(0,t), \frac{\partial}{\partial t} v(0,t)) \cdot \frac{d}{dt} v'(0)(t)\Big)dt\end{aligned}$$

and, in turn, by definition of the differential,

$$\begin{aligned}d\overline{\mathscr{L}_a^b}(v) &= \int_a^b \Big(\frac{\partial}{\partial q}\mathscr{L}(v(0,t), \frac{\partial}{\partial t} v(0,t)) \cdot v'(0)(t) \\ &\qquad - \frac{d}{dt}\frac{\partial}{\partial \dot{q}}\mathscr{L}(v(0,t), \frac{\partial}{\partial t} v(0,t)) \cdot v'(0)(t)\Big)dt \\ &\qquad + \frac{\partial}{\partial \dot{q}}\mathscr{L}(v(0,t), \frac{\partial v}{\partial t}(0,t))v'(0)(t)|_{t=a}^{t=b}.\end{aligned}$$

□

Denote

$$\widetilde{Q^{[a,b]}} = \| \, q \in Q^{[a,b]} \mid \dot{q}(a) = \dot{q}(b) = 0 \, \|$$

and consider, by restriction,

$$\overline{\mathscr{L}_a^b} \colon \widetilde{C^{[a,b]}} \longrightarrow R.$$

For $q \in \widetilde{Q^{[a,b]}}$, say that q is a solution of the Euler-Lagrange equations associated with the Lagrangian $\mathscr{L}$ if q satisfies

$$\frac{\partial}{\partial q}\mathscr{L}(q,\dot{q}) - \frac{d}{dt}\frac{\partial}{\partial \dot{q}}\mathscr{L}|_{[a,b]} = 0\,.$$

From Lemma 4.18 we immediately get, under the same assumptions, that the solutions of the Euler-Lagrange equation above are also critical points for the action integral associated with the Lagrangian in the time interval $[a,b]$—indeed, let $v = q_\star$ in Lemma 4.18.

Conversely, Lemma 4.13 gives—using all of the axioms of SDG now—that the critical points of $\overline{\mathscr{L}_a^b}$ are those $q \in \widetilde{Q^{[a,b]}}$ for which q is almost a solution of the Euler-Lagrange equations, in the sense that q satisfies

$$\neg\neg\Big(\frac{\partial}{\partial q}\mathscr{L}(q,\dot{q}) - \frac{d}{dt}\frac{\partial}{\partial \dot{q}}\mathscr{L}(q,\dot{q})|_{[a,b]} = 0\Big)\,.$$

Once again, an appeal to global sections applied to the topos $\mathscr{E}$ leads us to recover the familiar classical result from the result valid internally in $\mathscr{E}$.

Remark 4.19 As pointed out by Lawvere [72], the axiomatic theory SDG on $(\mathscr{E}, R)$ intended to formalize, in terms of simpler objects, the study of dynamical systems involving a state space $\mathbf{X}$ together with a vector field on it. The construction of the Lagrangian and that of the vector field on states involves an analysis of several sorts of forces, depending in turn on a specific construction of the configuration space $\mathbf{Q}$, which is usually realized as a given subspace

$$\mathbf{Q} \subset \mathbf{E}^{\mathbf{B}}$$

(of 'placements') where $\mathbf{E} = \mathbf{E}_3$ is the actual space and where $\mathbf{B}$ is the space of 'particles' of the material body in question. In particle mechanics, $\mathbf{B}$ is a finite discrete set, but in continuous mechanics it is usually a 3-dimensional manifold. One of the motivations of SDG was to formalize in simple terms the old idea that the theory of the infinite-dimensional $\mathbf{Q}$ with $\dim(\mathbf{B}) > 0$ should be similar to that of the particle case. This was also the motivation for [29]. In [72], one can find several interesting ideas for future uses of the synthetic (or axiomatic) approach in connection with Physics.

PART THREE

TOPOSES AND DIFFERENTIAL TOPOLOGY

In this third part we introduce and then make essential use of the intrinsic local and infinitesimal concepts available in any topos. By analogy with the synthetic theory of differential geometry, where jets of smooth maps are assumed representable by suitable infinitesimal objects, the axioms of synthetic differential topology express the representability of germs of smooth mappings, also by infinitesimal objects. However, whereas the nature of the infinitesimals of synthetic differential geometry are algebraic, those of synthetic differential topology are defined using the full force of the logic of a topos. In both cases, the use of Heyting (instead of Boolean) logic introduces an unexpected conceptual richness. To the basic axioms of synthetic differential topology we add a postulate of infinitesimal inversion and a postulate of the (logical) infinitesimal integration of vector fields. To these axioms and postulates we add others that shall be needed in the theory of stable mappings and their singularities to be dealt with in the fourth part of this book. The validity of all of these axioms and postulates in a single model will be shown in the fifth part of the book.

5
Local Concepts in SDG

In topos theory, the name 'topology' is used in a specific way to refer to a local operator on the subobjects classifier Ω of a given topos $\mathscr{E}$. Recall that for any topology j on a topos $\mathscr{E}$ there is defined a full subcategory $i\colon \mathrm{Sh}_j(\mathscr{E}) \hookrightarrow \mathscr{E}$ whose objects are the j-sheaves in $\mathscr{E}$. The importance of such a notion is mainly that the so constructed full subcategory is itself a topos and that the inclusion i is part of a geometric morphism $a \dashv i$, where a ('the associated sheaf functor') preserves finite limits. Thus, the idea of a Grothendieck topos $\mathscr{F}$ as a category of sheaves on a site $(\mathbf{C}, \mathbf{J})$ in *Set* is recovered by means of a topology j on the presheaf topos $\mathscr{E} = Set^{\mathbf{C}^{\mathrm{op}}}$, so that $\mathscr{F} = \mathrm{Sh}_j(\mathscr{E})$ for a topology (local operator) j on the subobjects classifier Ω. Grothendieck toposes are important in that they provide well adapted models for the synthetic theories of differential geometry and topology. There is, however, another notion of 'topology' that we need to consider in what follows, and that is a notion of 'topological structure' for a topos $\mathscr{E}$ [25]. In turn, this was prompted by the notion of intrinsic (or Penon) opens [96]. The 'intrinsic topological structure', available in any topos, has special properties which makes it particularly useful. For this reason it is important, when dealing with other topological structures with a classical origin, to determine under what conditions those are 'subintrinsic' and even better, when do they agree with the intrinsic one. We focus on two particular such topological structures on models of SDG —the Euclidean and the weak topological structures [25], [26].

5.1 The Intrinsic Topological Structure

Recall that a *frame* in a topos $\mathscr{E}$ is a partially ordered object L of $\mathscr{E}$ with arbitrary internal suprema and finite infima, satisfying the distributive law

$$H \wedge \bigvee_{i \in I} G_i = \bigvee_{i \in I} (H \wedge G_i).$$

By a *subframe* of a frame L it is meant any subobject $S \subset L$ that is closed under arbitrary suprema and finite infima and which is such that the maximal element of L belongs to S.

The collection $\mathscr{O}(X)$ of all open subsets of a topological space X is a frame with $\vee$ and $\wedge$ the usual operations of arbitrary union and finite intersection, 0 and 1 given respectively by the open subsets $\emptyset$ and X of X. It can be made into a Heyting algebra with implication $U \Rightarrow V$ given by $\neg U \vee V$, where $\neg U \subset X$ is the *interior* of its (set-theoretic) complement. In particular, the interior of the complement of any point $x \in U$ is an open subset of X whose union with U (which is an open subset of X) is the entire space X.

The subobjects classifier Ω of a topos $\mathscr{E}$ is an internal frame in $\mathscr{E}$ with the usual operations of union and intersection of subobjects, 0 and 1 given respectively by $\bot$ and $\top$, from which it follows that, for each object X of $\mathscr{E}$, the object Ω^X of $\mathscr{E}$ is also a frame in $\mathscr{E}$.

Definition 5.1 Let $\mathscr{E}$ be a topos.

(i) A *topological structure* on an object X of $\mathscr{E}$ is any object $S(X)$ in $\mathscr{E}$ which is a subframe of Ω^X. This data is said to induce a *topological structure* S on the topos $\mathscr{E}$ itself if every $f \in Y^X$ is continuous relative to it. Explicitly this means that for every X, Y objects of $\mathscr{E}$, and $U \in \Omega^Y$,

$$U \in S(Y) \Rightarrow f^{-1}(U) \in S(X).$$

(ii) A *basis* B for a topological structure S is given, for each object X of $\mathscr{E}$, by a subinflattice $B(X) \subset \Omega^X$, with $B(X) \subset S(X)$, so that $S(X)$ is generated by $B(X)$ in the sense that

$$U \in S(X) \Leftrightarrow \forall x \in U \exists V \in B(X)(x \in V \subset U)$$

holds, and if it is enough to test S-continuity on B, that is, if for all $f \in Y^X$ in $\mathscr{E}$,

$$U \in B(Y) \Rightarrow f^{-1}(U) \in S(X).$$

We say alternatively that $S(X)$ is the topological structure generated by the basis $B(X)$.

Remark 5.2 Let S be a topological structure on $\mathscr{E}$, $U \in S(X \times Y) \subset \Omega^{X \times Z}$ and, for each $z \in Z$, let

$$U_z = [[x \in X \mid (x,z) \in U]] .$$

It follows from the continuity assumption in the definition of a topological structure (Definition 5.1) that $U_z \in S(X)$. Therefore, under the identification $\Omega^{X \times Z} \simeq (\Omega^X)^Y$, there is an inclusion $S(X \times Z) \subset S(X)^Z$. Similarly, $S(X \times Z) \subset S(Z)^X$.

The inclusion $S(X \times Z) \subset S(X)^Z$ is always strict except when $S(X) = \Omega^X$. Indeed, if $S(X \times Z) = S(X)^Z$ then $S(Z) = S(1)^Z$. But $S(1) = \Omega$ since it is a sublocale of Ω. For a base, however, it is possible to have $B(X \times Z) = B(X)^Z$ for an arbitrary X, which means that the base B is *classifiable* in the sense that $B(X) = B^X$ for some $B \subset \Omega$ and so, necessarily, that $B = B(1)$. Such a B is then a subinflattice of Ω that is a classifier of those parts of X which are in $B(X)$. In terms of families, this says that an S-open part $U \subset X \times Z$ determines a Z-indexed family of S-open parts of X, as depicted in the diagram below, but that the converse is false unless $S(X) = \Omega^Z$.

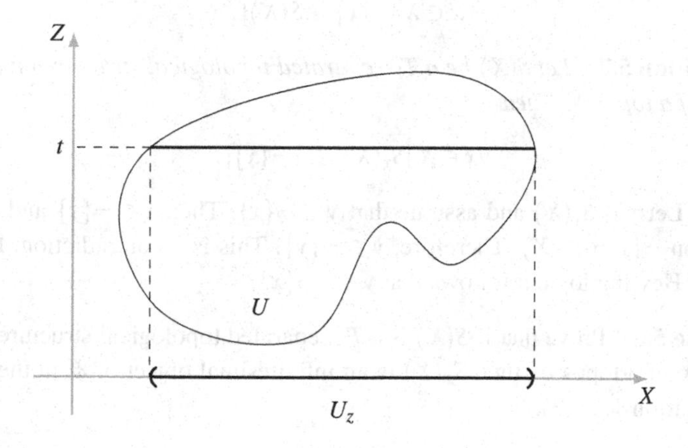

Figure 5.1 Open $U \subset X \times Z$

Definition 5.3 For any subframe $S \subset L$ of a frame L, there is an *interior operator*

$$l \colon L \longrightarrow L$$

defined so that for any $H \in L$, $l(H) \in S$ and $l(H) \subset H$ is the largest element of S smaller than H.

Definition 5.4 A topological structure $S(X)$ on an object X of a topos $\mathscr{E}$ is said to satisfy the *covering principle* if the following condition holds :

$$\forall H, G \in \Omega^X \, [H \cup G = X \Rightarrow l(H) \cup l(G) = X]$$

where l is the interior operator associated with S in the sense of Definition 5.3.

Given a topological structure $S(X)$ on an object X of a topos $\mathscr{E}$ and a point $x \in X$, denote by $S_x(X)$ the intersection of all S-open neighbourhoods of x in X, that is,

$$S_x(X) = \bigcap \{U \in S(X) \mid x \in U\}.$$

Remark 5.5 In a topological space in the topos *Set*, $S_x(X)$ reduces to a point under minimal separation properties for $S(X)$. For an arbitrary topos $\mathscr{E}$, such a condition in general merely implies that not every point of $S_x(X)$ is well separated from x. We shall be more precise in what follows.

Definition 5.6 A topological structure $S(X)$ on an object X of a topos $\mathscr{E}$ is said to satisfy the *separation condition* T_1 if in $\mathscr{E}$ holds

$$\forall x \in X \, [\neg\{x\} \in S(X)].$$

Proposition 5.7 *Let $S(X)$ be a T_1-separated topological structure on an object X of a topos $\mathscr{E}$. Then*

$$\forall x \in X \, [S_x(X) \subset \neg\neg\{x\}].$$

Proof Let $y \in S_x(X)$ and assume that $y \in \neg\{x\}$. Then $x \in \neg\{y\}$ and by assumption $\neg\{y\} \in S(X)$. Therefore, $y \in \neg\{y\}$. This is a contradiction. By the rules of Heyting logic it follows that $y \in \neg\neg\{x\}$. □

Exercise 5.8 Prove that if $S(X)$ is a T_1-separated topological structure on an object X of a topos $\mathscr{E}$, then $S_x(X)$ is an infinitesimal object of $\mathscr{E}$ in the sense of Definition 1.24 (3).

Just as in ordinary topology there is, along with a notion of a T_1-space for a topological structure $S(X)$, one of a T_2-space.

Definition 5.9 Given a topological structure S on a topos $\mathscr{E}$, an object X is said to be T_2-*separated* if, denoting by $\text{diag}_X \subset X \times X$ the diagonal,

$$\neg\text{diag}_X \in S(X \times X)$$

is satisfied.

Here is another notion which can be stated in the general case of a topos $\mathscr{E}$.

Definition 5.10 Let $S(X)$ be a topological structure on an object X of $\mathscr{E}$ and let H, G be two parts of X with $H \subset G$. We say that G *is an S-neighbourhood of* H if any of the two following equivalent conditions is satisfied:

(i) $$\forall x \in H \exists U \in S(X) (x \in U \subset G)$$

(ii) $$\exists U \in S(X) (H \subset U \subset X).$$

Under the same assumptions, we say that H is *well contained* in G if $\neg H \cup G = X$, that is, if the following condition is satisfied:

$$\forall x \in X (\neg(x \in H) \vee x \in G).$$

Proposition 5.11 *Let $S(X)$ be a topological structure in $\mathscr{E}$. Assume that $S(X)$ satisfies the covering principle. Then, if H is well contained in G, it follows that G is an S-neighbourhood of H.*

Proof Assume that H is well contained in G. Thus, $\neg H \cup G = X$. From the covering principle it follows that $\iota(\neg H) \cup G = X$, where ι is the interior operator corresponding to $S(X)$, and so also $\neg H \cup \iota(G) = X$. Then, $\forall x \in H\, x \in \iota(G)$. We have $l(G) \in S(X)$ and $H \subset \iota(G) \subset G$, so G is an S-neighbourhood of H. □

We proceed to discuss the intrinsic topological structure $P(X)$ on any object X of a topos $\mathscr{E}$. Recall for this the definition and properties of the intrinsic (or Penon) opens from Chapter 1, in particular Definition 1.20 and Remark 1.21.

Exercise 5.12 Prove using Proposition 1.22 that the singling out of the subframes $P(X)$ of Ω^X for objects X of a topos $\mathscr{E}$ gives a topological structure P on $\mathscr{E}$, to be referred to as the *intrinsic topological structure P*.

Remark 5.13 Since Ω has no proper subframes it follows that, in $\mathscr{E}$, $P(1) = \Omega$, so that every subobject of 1 is an intrinsic open.

Proposition 5.14 *Let $S(X)$ be a topological structure on an object X of a topos $\mathscr{E}$. Assume that it satisfies the covering principle in the sense of Definition 5.4. Then*

$$P(X) \subset S(X),$$

that is, every intrinsic or P-open is S-open.

Proof Let $U \in P(X)$ and let $x \in U$. By definition, $\neg\{x\} \cup U = X$. It follows from the covering principle for $S(X)$ that $\neg\{x\} \cup \iota(U) = X$. This implies that $x \in \iota(U)$ and so U is S-open. □

We shall say that a topological structure $S(X)$ is *subintrinsic* if every S-open is an intrinsic open, that is, if $S(X) \subset P(X)$.

Proposition 5.15 *If $S(X)$ is a subintrinsic T_1 topological structure on an object X of a topos $\mathscr{E}$, then for all $x \in X$,*

$$P_x(X) = S_x(X) = \neg\neg\{x\}.$$

Proof The claimed identities are a consequence of the chain of inclusions

$$S_x(X) \subset \neg\neg\{x\} \subset P_x(X) \subset S_x(X)$$

whose validity is next established. The first inclusion is a consequence of Proposition 5.7 since by assumption $S(X)$ is a T_1 topological structure. The second inclusion is a consequence of Remark 1.21. The third and last inclusion holds since $S(X)$ is subintrinsic, that is, since $S(X) \subset P(X)$ so that taking the intersections of all S-open neighbourhoods, respectively of all P-open neighbourhoods, of x, reverses the inclusion relation. □

Lemma 5.16 *Under the identifications*

$$\Omega^{X \times Y} = (\Omega^X)^Y = (\Omega^Y)^X,$$

one has the identification

$$P(X \times Y) = P(X)^Y \cap P(Y)^X.$$

Proof By Remark 5.2 applied to any topological structure, we already have

$$P(X \times Y) \subset P(X)^Y \cap P(Y)^X.$$

Our task is then to prove the converse. This means to prove that, if a part $U \subset X \times Y$ is such that for each $y_1 \in Y$ and all $x_1 \in X$ both $U_{y_1} = [[x \mid (x, y_1) \in U]]$ and $U_{x_1} = [[y \mid (x_1, y) \in U]]$ are intrinsic opens of X, respectively of Y, then $U \in P(X \times Y)$.

The two hypotheses can be translated, for $x_2 \in X$ and $y_2 \in Y$, into the statements

(i) $$(x_2, y_1) \in U \Rightarrow \forall y \in Y\ [\neg(y = y_1) \vee (x_2, y) \in U]$$

(ii) $$(x_1, y_2) \in U \Rightarrow \forall x \in X\ [\neg(x = x_1) \vee (x, y_2) \in U],$$

and to show

$$(x_0, y_0) \in U \Rightarrow \forall (x, y)\ [\neg((x, y) = (x_0, y_0)) \vee (x, y) \in U]$$

we do as follows. Let $(x_0, y_0) \in U$. Given $(x, y) \in X \times Y$, fix $y_0 = y_2$ in (2). Then,

$$\neg(x = x_0) \vee (x, y_0) \in U.$$

If $\neg(x = x_0)$, then $\neg((x,y) = (x_0,y_0))$ and we are done. If, on the other hand, $(x,y_0) \in U$ then let $x = x_2$ in (1). We then have

$$\neg(y = y_0) \vee (x,y) \in U .$$

If $\neg(y = y_0)$, then $\neg((x,y) = (x_0,y_0))$ and we are done. If, on the other hand, $(x,y) \in U$ we are done as well. This ends the proof. □

We next give a characterization of the T_2 condition for the intrinsic topological structure P on a topos $\mathscr{E}$.

Proposition 5.17 *Let X,Y be any two objects of $\mathscr{E}$ such that $P(X)$ and $P(Y)$ are both T_1. Then the following are equivalent:*

(i) *$P(X \times Y)$ is separated (that is, T_2).*
(ii) *For all $x_1,x_2 \in X$ and all $y_1,y_2 \in Y$,*

$$\neg((x_1,y_1) = (x_2,y_2)) \Rightarrow (\neg(x_1 = x_2) \vee \neg(y_1 = y_2)) .$$

Proof (1) $\Rightarrow$ (2). Clearly T_2 implies T_1, so $P(X \times Y)$ is T_1 since by assumption (1) it is T_2. Therefore,

$$\neg((x_1,y_1) = (x_2,y_2)) \Rightarrow \forall (x,y) \in X \times Y \left[\neg((x,y) = (x_1,y_1)) \right.$$
$$\left. \vee \neg((x,y) = (x_2,y_2)) \right] .$$

It suffices then to take $(x,y) = (x_1,y_2)$ in order to deduce (2).

(2) $\Rightarrow$ (1). By Lemma 5.16 applied to $(X \times Y) \times (X \times Y)$ it suffices to show that for all $(p,q) \in X \times Y$, $\neg\{(p,q)\} \in P(X \times Y)$. It is clear that (2) is actually an equivalence. Thus we have

$$\neg\{(p,q)\} = [[(x,y) \mid \neg((x,y) = (p,q))]]$$
$$= [[(x,y) \mid \neg(x = p)]] \cup [[(x,y) \mid \neg(y = q)]]$$

which is a union of two intrinsic opens of $X \times Y$, hence an intrinsic open. □

Corollary 5.18 *For any object X in a topos $\mathscr{E}$, $P(X)$ is T_2-separated if and only if it is T_1-separated.*

Proof Let $Y = 1$ in Proposition 5.17. Then $P(X)$ is T_2-separated if and only if for all $x_1,x_2 \in X$,

$$\neg(x_1 = x_2) \Rightarrow \forall x \in X [\neg(x = x_1) \vee \neg(x = x_2)] .$$

□

The intrinsic topological structure P on a topos $\mathscr{E}$ will not in general satisfy the covering principle in the sense of Definition 5.4. The following is a consequence of the covering principle for P when it does hold.

Proposition 5.19 *Let $\mathscr{E}$ be a topos and let $P(X)$ be the intrinsic topological structure on an object X of $\mathscr{E}$. Let $p \in X$ and $G \subset X$. Consider the conditions*

(i)
$$\neg\{p\} \cup G = X.$$

(ii)
$$\exists U \in P(X)\,(p \in U \subset G).$$

Then, under no further assumptions, (1) $\Rightarrow$ (2). *Moreover, this is also the case for any subintrinsic topological structure $S(X)$. If $P(X)$ satisfies the covering principle then* (2) $\Rightarrow$ (1), *hence* (1) *and* (2) *are equivalent conditions.*

Proof The first assertion is obvious. The second assertion follows from Proposition 5.11 applied to $S(X) = P(X)$. □

One can push this further. For any two objects X, Y of $\mathscr{E}$, and $x \in X$, let $H_x \subset Y$, $G_x \subset Y$ be any two parts of Y with $H_x \subset G_x$. Let $H \subset X \times Y$ be defined by $H = [[(x,y) \mid y \in H_x]]$ and let $G \subset X \times Y$ be given by $G = [[(x,y) \mid y \in G_x]]$. Clearly $H \subset G$. It is easy to see that $\neg H = [[(x,y) \mid y \in \neg(H_x)]]$. From this, in turn, follows the equivalence

$$\frac{\forall x \in X\ (\neg H_x \cup G_x = Y)}{\neg H \cup G = X \times Y}.$$

Corollary 5.20 *Let X and Y be objects of a topos $\mathscr{E}$ such that $P(X \times Y)$ satisfies the covering principle. Let $p \in Y$ and $H = X \times \{p\}$. Then, if $X \times \{p\} \subset G \subset X \times Y$, G is a P-neighbourhood of $X \times \{p\}$ in $X \times Y$ if and only if for each $x \in X$, G_x is a P-neighbourhood of $X \times \{p\}$ in $X \times Y$.*

Proof It follows from Proposition 5.19 using the above observation. □

In order to relate to the classical theory of differential topology, we need to interpret in our context two special topological structures – the Euclidean topological structure $E(R^n)$ and the weak topological structure $W(R^{mX})$, where $(\mathscr{E}, R)$ is a model of SDG.

5.2 The Euclidean and the Weak Topological Structures

We shall consider the intrinsic topological structure on the subobjects of R where $(\mathscr{E}, R)$ is a model of SDG.

Recall that $U \subset R$ is an intrinsic open (in the sense of Penon [96]) provided the following statement

$$\forall x \in U\, \forall y \in R\, [\neg(y = x) \vee y \in U]$$

holds in $\mathscr{E}$. For any object X of $\mathscr{E}$, denote by $P(X) \subset \Omega^X$ the subobject of intrinsic opens of X.

Proposition 1.22 implies in particular that

$$\forall f \in R^R \forall U \in \Omega^R \left[U \in P(R) \Rightarrow f^{-1}(U) \in P(R) \right].$$

By Postulate K (Postulate 2.10), R is a field in the sense of Kock, but it need not be a field in the usual sense. In particular, the object of the invertible elements of R, to wit

$$R^* = [[x \in R \mid \exists y \, (x \cdot y = 1)]],$$

is of interest as the following shows.

Proposition 5.21 *Let $(\mathscr{E}, R)$ be a model of SDG. Then*

$$R^* \in P(R).$$

Proof As shown in Proposition 2.12, it follows from Postulate K that R is a local ring. In particular, since for any $x, y \in R$, $x = (x - y) + y$, it follows that

$$\forall x \in R^* \forall y \in R \left[(x - y) \in R^* \vee y \in R^* \right].$$

Since $x - y \in R^* \subset \neg\{0\}$ in general, it follows that $\neg(y = x)$ hence $R^* \subset R$ is a Penon open. □

We are now in a position to introduce the Euclidean topological structure on R for any model $(\mathscr{E}, R)$ of SDG, taking as basic opens the open intervals $(a - \varepsilon, a + \varepsilon)$, $a \in R$. In R^n the basic opens will be the products of open intervals. Explicitly, given $x = (x_1, \ldots, x_n) \in R^n$ and $\varepsilon > 0$, let $B(x, \varepsilon) \subset R^n$ be the product of the intervals $(x_i - \varepsilon, x_i + \varepsilon)$, that is,

$$B(x, \varepsilon) = \bigcap_{i=1}^{n} [[y \in R^n \mid y_i - x_i \in (-\varepsilon, \varepsilon)]].$$

Then define the Euclidean topological structure $E(R^n)$ by letting

$$U \in E(R^n) \Leftrightarrow \forall x \in U \exists \varepsilon > 0 \, (B(x, \varepsilon) \subset U).$$

That $E(R^n) \subset \Omega^{R^n}$ is a subframe is clear since the only item to prove is that it is closed under finite infima, but this is a consequence of the second statement of Proposition 2.14.

There is no reason to limit ourselves to objects of the form R^n.

Definition 5.22 Given any object $M \subset R^n$ in $\mathscr{E}$, where $(\mathscr{E}, R)$ is a model of SDG, the Euclidean topology $E(M)$ is defined as the 'subspace' topology. Explicitly, for $U \in \Omega^M$,

$$U \in E(M) \Leftrightarrow \forall x \in U \, \exists \varepsilon > 0 \, (M \cap B(x,\varepsilon) \subset U).$$

We now compare the Euclidean with the intrinsic topological structures on any object M of $\mathscr{E}$ for a model $(\mathscr{E},R)$ of SDG.

Proposition 5.23 *Let $(\mathscr{E},R)$ be a model of SDG, and let M be any object of $\mathscr{E}$. Then, the Euclidean topological structure $E(M)$ is subintrinsic. That is, $E(M) \subset P(M)$.*

Proof It is enough to show that all open intervals $(x-\varepsilon, x+\varepsilon)$ are intrinsic or P-open. In fact, all open intervals (a,b) for $a,b \in R$ are P-open. We show this next. Let $x \in (a,b)$. To show that given any $z \in R$, the statement $\neg(z=x) \vee z \in (a,b)$ holds in $\mathscr{E}$. From (O3) follows that

$$z > a \vee z < x$$

as well as

$$z > x \vee b > z.$$

From the strictness of the order, that is, from (O2), we get the desired result as there are four possibilities three of which imply that $\neg(z=x)$ and the fourth that $z \in (a,b)$. □

Remark 5.24 Another consequence of Postulate O for a model $(\mathscr{E},R)$ of SDG is the following property of R:

$$\forall x \in R \neg\neg\{x\} = \bigcap_{\varepsilon>0} (x-\varepsilon, x+\varepsilon)$$

where $\neg\neg\{x\} = [[y \in R \mid \neg\neg(y=x)]]$. Recall that intuitively, to assert $\neg(y=x)$ is to express that y is well separated from x, as Fig. 5.2 shows. In the same picture one can visualize $\neg\neg\{x\}$ as the object consisting of those elements of R which are not well separated from x. Since the identity $\neg\{x\} \cup \neg\neg\{x\} = R$ is not valid, there is a part of R that has no explicit characterization in the topos, and it is often referred to as a *no man's land*. This part is depicted in white in the picture.

The assumptions made on R are strong enough to imply the T_2 (and T_1) separation conditions on the Euclidean topological structure on R^n.

Recall that, for any given topological structure S on a topos $\mathscr{E}$, $S_x(X)$ denotes the monad of $x \in X$. If $(\mathscr{E},R)$ is a model of SDG, we have $P_0(R^n)$ and $E_0(R^n)$ as well as $\Delta(n) = \neg\neg\{0\}$ for $0 \in R^n$. Since the Euclidean topological structure is subintrinsic, it follows (by taking intersections) that $P_0(R^n) \subset E_0(R^n)$.

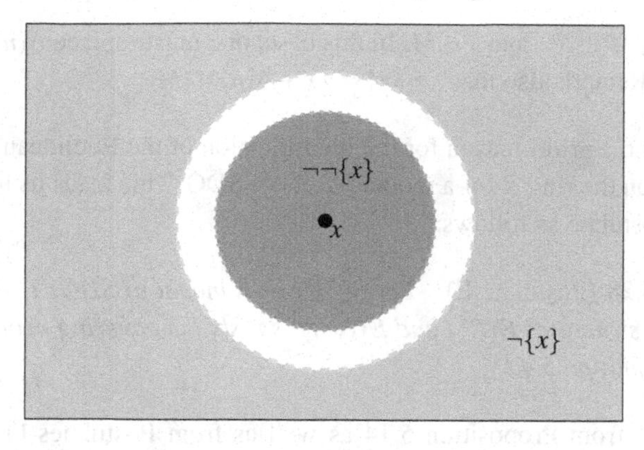

Figure 5.2 Monad of a point

Remark 5.25 In general the intrinsic topological structure on an object of a topos need not satisfy the separation conditions T_1 or T_2. However, the situation is better if $(\mathscr{E}, R)$ is a model of SDG. On account of the compatibility of the order relation with the ring structure it is enough to establish that $P(R)$ is T_1 at the origin, that is, to establish that

$$\forall x \in R\,[\neg(x=0) \Rightarrow (x>0 \vee x<0)],$$

that is that $\neg\{0\} = (\infty, 0) \cup (0, \infty)$. This is immediate from (O4). Thus, by translation, $P(R)$ is T_1-separated. It is also T_2-separated, as in the classical argument for it. Indeed, let $(x,y) \in \neg\mathrm{diag}_R$. Say that $\neg(x=0)$. Then by T_1-separated, it follows that either $x>y$ or $x<y$. In the first case, $z = \frac{x+y}{s}$ satisfies $x<z<y$, so $(x,y) \in (-\infty, z) \times (z, +\infty)$ and also $(-\infty, z) \times (z, +\infty) \subset \neg\mathrm{diag}_R$. A similar argument applies to the alternative situation. We have used the case $n=1$ of (O4) , but it is clear how the same argument works for R^n for an arbitrary n using (O4) in the same manner.

Proposition 5.26 *Let $(\mathscr{E}, R)$ be a model of SDG. Then the following hold.:*

(i) *The Euclidean topological structure $E(R^n)$ is T_2-separated for each positive integer n.*
(ii) *$E(R^n) \subset P(R^n)$ (thus $P(R^n)$ is also T_2-separated).*
(iii) *$P_0(R^n) = E_0(R^n) = \Delta(n)$.*

Proof It follows from Proposition 5.23, Remark 5.25, and Proposition 5.15. □

Remark 5.27 It is straightforward to check that Proposition 5.26 holds for any M in $\mathscr{E}$, $M \subset R^n$, and $x \in M$. In this case, one must replace $\Delta(n) = \neg\neg\{0\}$ by $\neg\neg\{x\}$. Remark also that $\neg\neg\{x\} = (x + \Delta(n)) \cap M$.

There is no a priori reason for the identification of the Euclidean and Penon topologies on the ring R for a model $(\mathscr{E}, R)$ of SDG. This leads us to introduce a special postulate as follows.

Postulate 5.28 (Postulate E) *Let $(\mathscr{E}, R)$ be a model of SDG. The Euclidean topological structures $E(R^n)$ and $E(\Delta(n))$ satisfy the covering principle in the sense of Definition 5.4.*

It follows from Proposition 5.14 as well as from Postulates O and E that $P(R^n) = E(R^n)$. In particular, both topological structures are T_2-separated and satisfy the covering principle. We can then assert that every $f\colon R^n \longrightarrow R^n$ is internally continuous in the Euclidean topological structure since this is the case for the intrinsic topological structure.

We now turn to the weak topological structure on functionals, that is, in our context, on objects of the form R^{nX}. The idea is to internalize the weak C^∞-topology used by G. Wassermann [110] for objects of the form R^{nX}.

Classically, the weak topology on $C^\infty(\mathbb{R}^n)$ has as a basis the sets

$$V(K, r, g, U) = \{h \in C^\infty(\mathbb{R}^n) \mid J^r(g - h)K \subset U\},$$

where $K \subset \mathbb{R}^n$ is a compact subset, $g \in C^\infty(\mathbb{R}^n)$, $0 \leq r \leq n$, $\varepsilon \in \mathbb{R}$, $\varepsilon > 0$, and $J^r f$ denotes the r-jet of $f \in C^\infty(\mathbb{R}^n)$. In terms of sequences, this topology is characterized by the following property: a sequence $\{f_n\}$ of smooth mappings is said to converge to a smooth mapping f in the weak C^∞-topology if it converges uniformly to it on any compact subset, and if the same holds for the sequence of all its derivatives. On $C_0^\infty(\mathbb{R}^n)$, the weak topology is the quotient topology.

In what follows we shall assume, in addition to Axioms J and W for $(\mathscr{E}, R)$, that the latter is also a model of Postulate O and Postulate E. Recall the definition of an object of a topos being compact in the sense of [36], chosen so to retain a property which any compact space K has —to wit, given any point x_0 in any topological space X, and a neighbourhood H of the fibre $\pi^{-1}(x_0) \subset K \times X \xrightarrow{\pi} X$, then there exists a neighbourhood U of x_0 such that $\pi^{-1}(U) \subset H$.

Definition 5.29 An object K of a topos $\mathscr{E}$ is said to be *compact* if the following holds:

$$\forall A \in \Omega\, \forall B \in \Omega^K\, [\forall k \in K\, (A \vee B(k)) \Rightarrow A \vee \forall k \in K\, B(k)].$$

The internal interpretation of this formula reads as follows

$$\forall A \in \Omega \, \forall B \in \Omega^K \, [\forall \pi_k \, (\pi_K^{-1} A \cup B) \subset A \cup \forall_{\pi_K} B] \, ,$$

or, equivalently,

$$\forall A \in \Omega \, \forall B \in \Omega^K \, [K = \pi_K^{-1} A \cup B \Rightarrow \mathbf{1} = A \cup \forall_{\pi_K} B],$$

where $\pi_K : K \longrightarrow \mathbf{1}$ is the unique morphism to the terminal object.

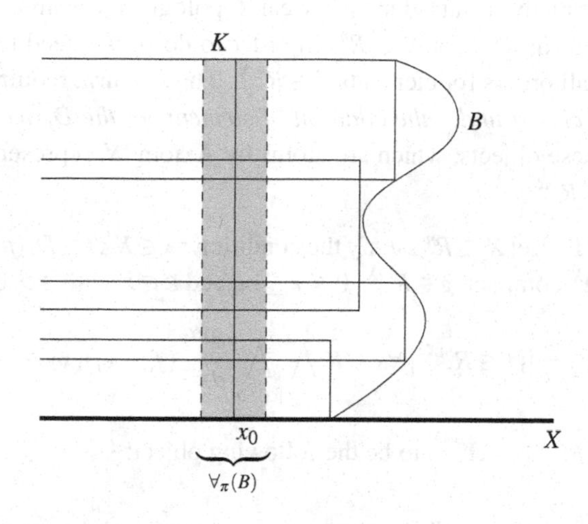

Figure 5.3 Compact object

Proposition 5.30 *Let X be any object of $\mathscr{E}$. Then the following holds:*

$$\forall K, L \in \Omega^X \, [K \text{ compact} \wedge L \text{ compact} \Rightarrow K \cup L \text{ compact}] \, .$$

Proof The derivations below are valid and constitute a proof of the statement.

$$\frac{\dfrac{\dfrac{\dfrac{K \cup L = \pi_{K \cup L}^{-1} A \cup B}{K = \pi_K^{-1} A \cup (K \cap B) \wedge L = \pi_L^{-1} A \cup (L \cap B) \ (*)}}{\mathbf{1} = A \cup \forall_{\pi_K} (K \cap B) \wedge \mathbf{1} = A \cup \forall_{\pi_L} (L \cap B)}}{\mathbf{1} = A \cup (\forall_{\pi_K} (K \cap B) \cap \forall_{\pi_L} (L \cap B)) \ (**)}}{\mathbf{1} = A \cup \forall_{\pi_{K \cup L}} (B)}$$

It remains to verify $(*)$ and $(**)$. From the two inclusions $u_K : K \hookrightarrow K \cup L$ and $u_L : L \hookrightarrow K \cup L$ we clearly have $\pi_{K \cup L} \circ u_K = \pi_K$ and $\pi_{K \cup L} \circ u_L = \pi_L$. To get $(*)$ intersect with K first, then with L and apply the above. To get $(**)$ it is enough to notice that, on account of the above identities, one has that

$$\forall_{\pi_K}(K \cap B) = \forall_{\pi_{K \cup L}}(\forall_{\pi_K}(K \cap B)) = \forall_{\pi_{K \cup L}}(B),$$

and similarly for L. □

We are now ready to introduce the weak topological structure on function objects of the form R^{mX} for $X \subset R^n$. In order to do so we need to use partial derivatives of all orders for elements $f \in R^{mX}$. This, in turn, requires that such an $X \subset R^n$ be *closed under the addition of elements of the* $D_r(n) \subset R^n$ since, by Axiom J these objects, which are atoms by Axiom W, represent r-jets at 0 of elements of R^{nR^m}.

Definition 5.31 Let $X \subset R^n$ satisfy the condition $\forall x \in X \forall t \in D_r(n) [x + t \in X]$. For any $K \in \Omega^X$ compact, $g \in R^{mX}$, $0 \leq r \leq m$, and $\varepsilon \in R$ with $\varepsilon > 0$, we denote

$$V(K, r, g, \varepsilon) = [[f \in R^{mX} \mid \forall x \in K \bigwedge_{i=1}^{n} \bigwedge_{|\alpha| \leq r} \frac{\partial^{|\alpha|}}{\partial x^{\alpha}}(f_i - g_i)(x) \in (-\varepsilon, \varepsilon)]]$$

and define $W(R^{mX}) \subset \Omega^{R^{mX}}$ to be the following object:

$$[[U \in \Omega^{R^{mX}} \mid \forall g \in U \exists K \in \Omega^X \exists \varepsilon \in R (K \text{compact} \wedge \varepsilon > 0 \wedge \bigvee_{r=0}^{n} V(K, r, g, \varepsilon) \subset U)]].$$

Proposition 5.32 *For any $n > 0$ and $X \subset R^n$ closed under the addition of elements of the $D_r(n)$, $W(R^{mX})$ is a topological structure, that is $W(R^{mX}) \subset \Omega^{R^{mX}}$ is a subframe.*

Proof We need only determine closure of the basic opens under infima. Given $K, L \in \Omega^X$, K, L compact, $0 \leq r, s \leq n$, $\varepsilon, \delta > 0$ and $g \in R^{mX}$, notice that

$$V(K \cup L, t, g, \gamma) \subset V(K, r, g, \varepsilon) \cap V(L, s, g, \delta)$$

where $t = \max(r, s)$ and $\gamma > 0$ exists by Proposition 2.14 (ii). That $K \cup L$ is compact follows from Proposition 5.30. □

The following general result for the weak topological structure will be useful when dealing with stability of germs in the synthetic context.

Proposition 5.33 *Let $n > 0$ and $X \subset R^n$ be closed under the addition of elements of the $D_r(n)$. Then the weak topological structure on R^{mX} is subintrinsic, that is, $W(R^{mX}) \subset P(R^{mX})$.*

Proof It is enough to show that for any $K \in \Omega^X$, K compact, $0 \leq r \leq n$, $\varepsilon \in R$, $\varepsilon > 0$ and $0 \in R^{mX}$,

$$V(K,r,0,\varepsilon) \in P(R^{mX}).$$

Recall that

$$V(K,r,0,\varepsilon) = \bigcap_{|\alpha|\leq r} \left(\frac{\partial^{|\alpha|}}{\partial x^{\alpha}}\right)^{-1} [[f \in R^{mX} \mid \forall x \in K \bigwedge_{i=1}^{m} (f_i(x) \in (-\varepsilon,\varepsilon))]].$$

By continuity of the intrinsic topological structure it is enough to show that the

$$Y(K,\varepsilon) = [[f \in R^{mX} \mid \forall x \in K \bigwedge_{i=1}^{m} (f_i(x) \in (-\varepsilon,\varepsilon))]]$$

are intrinsic open.

We have the following valid deduction.

$$\frac{\forall x \in K \;\; [h = f \Rightarrow \bigwedge_{i=1}^{m} (h_i(x) = f_i(x))]}{\forall x \in K \;\; [\neg \bigwedge_{i=1}^{m} (h_i(x) = f_i(x)) \Rightarrow \neg (h = f)]}.$$

Since $(-\varepsilon,\varepsilon)$ is E-open, it is also P-open, so we always have

$$\forall h \in R^{mX} \forall f \in Y(K,\varepsilon) \forall x \in K \;\; \bigwedge_{i=1}^{m} [\neg (h_i(x) = f_i(x)) \vee h_i(x) \in (-\varepsilon,\varepsilon)]$$

from which it follows intuitionistically that

$$\forall h \in R^{mX} \forall f \in Y(K,\varepsilon) \forall x \in K \;\; [\neg \bigwedge_{i=1}^{m} (h_i(x) = f_i(x)) \vee \bigwedge_{i=1}^{m} (h_i(x) \in (-\varepsilon,\varepsilon))].$$

We also have the valid deduction

$$(h = f) \Rightarrow \forall x \in K \bigwedge_{i=1}^{m} (h_i(x) = f_i(x)).$$

By the previous observation, it follows then that *a fortiori*

$$\forall h \in R^{mX} \forall f \in Y(K,\varepsilon) \forall x \in K \;\; [\neg (h = f) \vee \bigwedge_{i=1}^{m} (h_i(x) \in (-\varepsilon,\varepsilon))]$$

and, by compactness of K,

$$\forall h \in R^{mX} \forall f \in Y(K,\varepsilon) \;\; [\neg (h = f) \vee \forall x \in K \bigwedge_{i=1}^{m} (h_i(x) \in (-\varepsilon,\varepsilon))],$$

that is,

$$\forall h \in R^{mX} \forall f \in Y(K,\varepsilon) \;\; [\neg (h = f) \vee h \in Y(K,\varepsilon)].$$

This shows that $Y(K,\varepsilon)$ is intrinsic open and finishes the proof. □

Remark 5.34 In the next section we shall state axioms for synthetic differential topology in which the object $\Delta(n) = \neg\neg\{0\} \in R^n$ will play a special role. In anticipation for it we end this section with results that will be relevant therein.

Proposition 5.35 *$\Delta \subset R$ is closed under addition.*

Proof We wish to prove

$$\forall s,t \in R \ [(\neg\neg(s=0) \wedge \neg\neg(t=0)) \Rightarrow \neg\neg(s+t=0)].$$

Since R is a field in the sense of Kock (by Postulate K), we have

$$\forall s,t \in R \ [\neg(s+t=0) \Rightarrow (\neg(s=0) \vee \neg(t=0))].$$

Using Heyting logic we have that

$$(\neg(s=0) \vee \neg(t=0)) \Rightarrow \neg((s=0) \wedge (t=0))$$

and, given that $\neg\neg((s=0) \wedge (t=0))$ and $\neg\neg(s=0) \wedge \neg\neg(t=0))$ are equivalent in Heyting logic, the result follows by contraposition. □

Remark 5.36 (i) In the proof of Proposition 5.35 it should be noted that, although *reductio ad absurdum* is not in general valid in Heyting logic, it is so when applied to formulas of the form $\varphi \Rightarrow \neg\psi$ by getting a contradiction from $\varphi \wedge \psi$.

(ii) The last paragraph in the proof of Proposition 5.35 relies on Exercise 1.18.

Proposition 5.37 *For any $n > 0$, $\Delta(n) \subset R^n$ is compact.*

Proof Let $A \in \Omega$, $B \in \Omega^{\Delta(n)}$, $\pi \colon \Delta(n) \to \mathbf{1}$ the unique morphism into the terminal object, which is an epimorphism on account of the existence of a global section

$$\mathbf{1} \xrightarrow{0} \Delta(n).$$

Assume that

$$\Delta(n) = \pi^{-1}A \cup B.$$

By the covering principle, which holds for the intrinsic topological structure because it holds for the Euclidean topological structure on $\Delta(n)$, we have

$$\Delta(n) = \iota(\pi^{-1}A) \cup \iota(B)$$

where ι is the interior operator corresponding to the intrinsic topological structure P. Since $P(\Delta(n))$ is trivial, if $0 \in \iota(\pi^{-1}A)$, then $\pi^{-1}A = \Delta(n)$, and if

$0 \in \iota(B)$ then $B = \Delta(n)$. In the first case, in the pullback

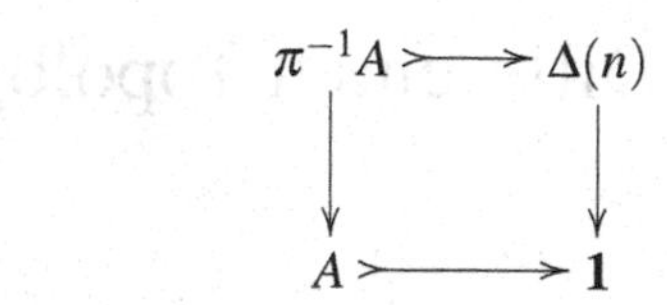

the top arrow is an iso hence an epimorphism, so that its composite with $\Delta(n) \twoheadrightarrow 1$ is also an epimorphism from which it follows that the bottom arrow $A \rightarrowtail\!\!\!\!\rightarrow 1$ is an epimorphism hence an isomorphism. In the second case $\forall_\pi B = \mathbf{1}$. In either case the conclusion is that

$$\mathbf{1} = A \cup \forall_\pi B.$$

□

Remark 5.38 (i) Among the possible applications of Proposition 5.35, in addition to $W(R^{R^n})$, is to $W(R^{\Delta(n)})$. This will be useful in connection with an axiom of synthetic differential topology (to be stated) that declares the latter to be the object of germs at 0 of maps $R^n \longrightarrow R$.

(ii) A further simplification arises from the fact that the topological structure $W(R^{\Delta(n)})$ may be defined by considering a single type of basic W-opens, to wit, those of the form $V(\Delta(n), r, 0, \varepsilon)$ on account of Proposition 5.37.

6

Synthetic Differential Topology

A subject which we shall call SDT (Synthetic Differential Topology) is here formally introduced by adding axioms of a local nature to SDG (Synthetic Differential Geometry). The appearance of [96], and in particular the consideration of the logical infinitesimal $\Delta = \neg\neg\{0\} \subset R$, where R is the ring of line type in a model $\mathscr{E}$ of SDG, opened up the way of a synthetic approach to a theory of germs of smooth maps by analogy with the theory of jets in SDG [61]. The intuitive idea of J. Penon [96] that germs of smooth maps $R^n \longrightarrow R^m$ should be thought of as represented by the logical infinitesimal $\Delta(n)$ was given the status of an axiom in [25]. Also in [25], the integrability of germs at 0 of maps $R^n \longrightarrow R$ up to $R^n \times \Delta$ was stated as a postulate. Among the postulates of SDT are those of infinitesimal inversion [96] and of density of regular values [26]. Several aspects of a theory of germs within SDT can be found in [20], [44], [26] and [103], and are included in this chapter.

6.1 Basic Axioms and Postulates of SDT

Let $(\mathscr{E}, R)$ be a model of SDG where $\mathscr{E}$ is a topos with a natural numbers object. Denote by

$$C_0^g(R^n, R^m)$$

the object in $\mathscr{E}$ of *germs at* 0 *of maps* $R^n \longrightarrow R$, where by the latter it is meant an equivalence class of elements $f \in \text{Partial}(R^n, R^m)$ with domain $\partial(f)$ such that $0 \in \partial(f) \in P(R^n)$, and where the equivalence relation for $f, l \in \text{Partial}(R^n, R^m)$ is given as follows:

$$f \sim l \Leftrightarrow \exists U \in P(R^n)\left(0 \in U \subset \partial(f) \cap \partial(l) \wedge f|_U = l|_U\right),$$

defined in the internal logic of the topos $\mathscr{E}$. Since $\Delta(n) = \neg\neg\{0\}$ is the intersection of all intrinsic opens of R^n, it follows that

$$\Delta(n) \subset \partial(f)$$

for any representative f of a germ. There is therefore a map

$$C_0^g(R^n, R^m) \xrightarrow{j} R^{m\Delta(n)},$$

given by restriction. The first axiom of SDT asserts the representability of germs. It brings a considerable simplification to the entire enterprise.

For future use we state it for the more general situation where the domain of a germ from R^n to R^m is an intrinsic open $M \subset R^n$ such that $0 \in M$.

Axiom 6.1 (Axiom G) *Let $0 \in M \in P(R^n)$, where $(\mathscr{E}, R)$ is a model of SDG. The restriction map*

$$j \colon C_0^g(M, R^m) \longrightarrow R^{m\Delta(n)}$$

is an isomorphism.

That the 'monads' of the type $\neg\neg\{0\}$ ought to be tiny is an assumption waiting to be fully explored. There is no harm in postulating it in our theory since, as we shall see, it is consistent with all other axioms and postulates of what we shall understand here by SDT.

Axiom 6.2 (Axiom M) *For any $n > 0$, the object $\Delta(n) = \neg\neg\{0\}$ of $\mathscr{E}$, with $0 \in R^n$, is an atom, that is, the endofunctor*

$$(-)^{\Delta(n)} \colon \mathscr{E} \longrightarrow \mathscr{E}$$

has a right adjoint[1].

The subobject $\Delta(n) \subset R^n$ represents germs at 0 of mappings from R^n to R directly, rather than by the quotient topology. It is for that reason that the properties established in the previous chapter in connection with the weak topological structure on objects of the form $R^{m\Delta(n)}$ acquire here a special significance. More precisely, it follows from Propositions 5.35 and 5.37 that, in the case of function spaces of germs of smooth mappings in SDT, the weak topological structure $W(R^{m\Delta(n)})$ needs consideration of just a single type of basic W-open, to wit, the $V(\Delta(n), r, f, \varepsilon)$. This is a substantial simplification, not available in the classical setting.

In what follows we assume that $(\mathscr{E}, R)$ is a model of SDG (Axioms J and W,

[1] Since $\Delta(n)$ has a global section, to wit $\ulcorner 0 \urcorner \colon 1 \longrightarrow \Delta(n)$, it is a well supported object—a condition which, added to that of atom, states that it is a tiny object. In fact, $\Delta(n)$ is an infinitesimal object.

Postulate K, Postulate F, Postulate O) that in addition satisfies Axioms G and M stated above. We refer to such an $(\mathscr{E}, R)$ as a *basic model of SDT*. Further postulates will be added in this chapter after some preliminaries.

Recall that in any model $(\mathscr{E}, R)$ of SDG there is defined the Euclidean topological structure $E(M)$ on any subobject $M \subset R^n$ for any $n > 0$. In this generality, it was shown that the Euclidean topology is subintrinsic—that is, $E(M) \subset P(M)$, where P is the intrinsic (or Penon) topological structure on $\mathscr{E}$ as a topos.

We next consider a suitable assumption to make about the existence and uniqueness of solutions to ordinary differential equations after some exploratory considerations.

Given any intrinsic open $M \subset R^m$ for some $m > 0$, such that $0 \in M$, some $g \in R^{mM}$ and a point $x \in M$, there is determined a differential equation

$$y' = g(y), \ y(0) = x.$$

A solution to this equation is a map f on the variables $(x,t) \in M \times R$ such that

$$\frac{\partial f}{\partial t}(x,t) = g(f(x,t)), \ f(x,0) = x.$$

It will have as domain of definition some intrinsic open H of $M \times R$ with

$$M \times \{0\} \subset H.$$

Remark 6.3 In any model of SDG, for $M = R^m$, a solution f of a differential equation arising from some $g \in R^{mM}$ with $H = M \times D_\infty$ exists and is unique. However, for $M = R^m$, or in the more general case of an intrinsic open M of R^m containing 0, there is no *a priori* reason why a solution on $H = M \times \Delta$ should (uniquely) exist. For this reason, the latter was assumed for the case $M = R^m$ in [24] and named Postulate WA2. We shall actually need a more general version of it which is convenient for deducing the existence and uniqueness of solutions of time-dependent systems of ordinary differential equations. Before stating it as Postulate S, we first recall Postulate WA2 in a suitably modified form, as follows.

Postulate 6.4 (Postulate WA2) *Let $(\mathscr{E}, R)$ be a basic model of SDT. Let $m > 0$. Let $0 \in M \in P(R^m)$. Then the following is postulated.*

$$\forall g \in R^{mM} \exists ! f \in M^{M \times \Delta} \forall x \in M \forall t \in \Delta$$
$$\left[f(x,0) = x \wedge \frac{\partial f}{\partial t}(x,t) = g(f(x,t)) \right].$$

In this form, Postulate WA2 may not seem to correspond to the classical version. However, Axiom G provides a way to pass from the infinitesimal to the local as the next proposition shows.

Proposition 6.5 *Given a flow $f \in M^{M\times\Delta}$ to a certain vector field $g \in R^{mM}$, where $0 \in M \in P(R^m)$, there exists (uniquely) a local flow that extends f. The uniqueness means that any two such extensions agree on an intrinsic open neighbourhood of $M \times \{0\}$.*

Proof It follows from Axiom G that there are extensions $h \in M^U$ with

$$M \times \{0\} \subset U \in P(M) \times R$$

of f and that any two such extensions agree on an intrinsic open neighbourhood of $M \times \{0\}$. It remains to verify the flow equation. Let

$$H = [[(x,t,r) \mid (x,t+r) \in U \wedge (h(x,t),r) \in U]].$$

Clearly

$$M \times \{0\} \subset H \subset M \times R^2$$

and

$$H \in P(M \times R^2).$$

Let h_1 and h_2 defined on H be given by

$$h_1(x,t,r) = h(x,t+r)$$

and

$$h_2(x,t,r) = h(h(x,t),r).$$

Since f is a flow and h extends f, h_1 and h_2 agree on $M \times \Delta(2)$. Then, again by Axiom G, they also agree on an intrinsic open $W \in P(M \times R^2)$ with

$$M \times \{0\} \subset W \subset H.$$

Since addition is an open map for the Euclidean topological structure (trivial verification) and since $P = E$, it follows that h satisfies the flow equation on a neighbourhood (possibly smaller than U) of $M \times \{0\}$. □

Theorem 6.6 *Let $(\mathscr{E},R)$ be a basic model of SDT which also satisfies Postulate WA2. Let $m > 0$. Let $0 \in M \in P(R^m)$. Then the following holds.*

$$\forall g \in R^{mM} \exists! f \in M^{M\times\Delta} \forall x \in M \forall d \in D \Big[f(x,d) = x + d \cdot g(x) \\ \wedge \forall t,r \in \Delta\big(f(x,t+r) = f(f(x,t),r)\big)\Big].$$

Proof That the statement is meaningful is a consequence of Proposition 5.35. The first statement inside the square brackets is a consequence of Axiom J. The second statement inside the square brackets is a consequence of the uniqueness part of Postulate WA2. If f is a solution to the differential equation determined by g with initial value x, then for each $(x,t) \in M \times \Delta$ both functions $y_1(t,r) = f(x,t+r)$ and $y_2 = f(f(x,t),r)$ satisfy the differential equation $y' = g(y)$ with initial condition $y(0) = f(x,0)$, hence they must agree on an open contained in their common parts of definition. □

Remark 6.7 The statements of Postulate 6.4 and Theorem 6.6 are equivalent. Indeed, letting $t \in \Delta$ and $d \in D$ (so in particular $d \in \Delta$), the flow equation condition in Theorem 6.6 reduces to

$$\forall t \in \Delta \forall d \in D\big(f(x,t+d) = f(f(x,t),d)\big).$$

Using Axiom J, we obtain

$$f(x,t+d) = f(f(x,t),d) = f(x,t) + d \cdot g(f(x,t))$$

and

$$f(x,0+0) = f(f(x,0),0) = f(x,0) = x.$$

The following is a generalization of Postulate WA2 that will be used in connection with solutions to time-dependent systems of differential equations.

Postulate 6.8 (Postulate S) *Let $(\mathscr{E},R)$ be a basic model of SDT. Let $m > 0$. Let $0 \in M \in P(R^m)$. Then the following is postulated.*

$$\forall g \in (R^m \times [0,1])^{M\times[0,1]} \exists! f \in (M \times [0,1])^{M\times[0,1]\times\Delta} \forall x \in M \forall s \in [0,1] \forall t \in \Delta \left[f(x,s,0) = (x,s) \wedge \frac{\partial f}{\partial t}(x,s,t) = g(f(x,s,t))\right].$$

Exercise 6.9 (i) Deduce Postulate WA2 from Postulate S.

(ii) Prove that Postulate S can be equivalently stated as follows.

$$\forall g \in (R^m \times [0,1])^{M\times[0,1]} \exists! f \in (M \times [0,1])^{M\times[0,1]\times\Delta} \forall x \in M \forall s \in [0,1] \forall d \in D \left[f(x,s,d) = (x,s) + d \cdot g(x,s) \wedge \forall t,r \in \Delta\big(f(x,s,t+r) = f(f(x,s,t),r)\big)\right].$$

We shall make use of item 2 of Exercise 6.9 in the proof of the following proposition about time-dependent systems.

Proposition 6.10 *Let $(\mathscr{E},R)$ be a basic model of SDT which satisfies Postulate S. Let $m > 0$. Let $0 \in M \in P(R^m)$. Then the following holds:*

$$\forall g \in R^{mM\times[0,1]} \exists! f \in M^{M\times\Delta} \forall x \in M \forall t \in \Delta \left[f(x,0) = x \wedge \frac{\partial f}{\partial t}(x,t) = g(f(x,t),t)\right].$$

Proof Let

$$\hat{g} \in (R^m \times [0,1])^{M\times[0,1]}$$

be given by

$$\hat{g}(x,s) = (g(x,s),1).$$

By Postulate S there exists a unique

$$\hat{f} \in M \times [0,1]^{M\times[0,1]\times\Delta}$$

such that

$$\forall x \in M \forall s \in [0,1] \hat{f}(x,s,0) = (x,s) \wedge \forall t \in \Delta \frac{\partial \hat{f}}{\partial t}(x,s,t) = \hat{g}(\hat{f}(x,s,t)).$$

Equivalently (as in Exercise 6.9), for all $(x,s) \in M \times [0,1]$, $d \in D$, $t, r \in \Delta$ we have (**)

$$\hat{f}(x,s,d) = (x,s) + d \cdot \hat{g}(x,s)$$

and

$$\hat{f}(x,s,t+r) = \hat{f}(\hat{f}(x,s,t),r).$$

Letting $\hat{f}(x,s,t) = (\hat{f}_1(x,s,t), \hat{f}_2(x,s,t))$ so that

$$\hat{f}_1 \in M^{M\times[0,1]\times\Delta}$$

and

$$\hat{f}_2 \in [0,1]^{M\times[0,1]\times\Delta},$$

we get two separate sets of conditions as follows. First, since the unique extension of the map $D \longrightarrow R$ given by $d \mapsto s+d$ for a fixed $s \in [0,1]$ is a Δ-flow, it must be the map $t \mapsto s+t \colon \Delta \longrightarrow R$. In other words, we must have $(**)_1$

$$\hat{f}_2(x,s,t) = s+t$$

for all $x \in R^m$, $s \in [0,1]$, $t \in \Delta$. With it, the second equation above reduces to the (uninteresting) equation

$$s+(t+r) = (s+t)+r.$$

We may also use the fact that $\hat{f}_2(x,s,t) = s+t$ in the following $(**)_2$, also derived from $(**)$:

$$\hat{f}_1(x,s,d) = x + d \cdot g_1(x,s)$$

and

$$\hat{f}_1(x,s,t+r) = \hat{f}_1(\hat{f}_1(x,s,t),s,r)$$

or, equivalently,

$$\hat{f}_1(x,s,0) = x$$

and

$$\frac{\partial \hat{f}_1}{\partial t}(x,s,t) = g(\hat{f}_1(x,s,t), s+t).$$

Finally, from $\hat{f}_1 \in M^{M\times[0,1]\times\Delta}$, obtain $f \in M^{M\times\Delta}$ given as

$$f(x,t) = \hat{f}_1(x,0,t).$$

This f is unique satisfying

$$f(x,0) = x$$

for all $x \in M$ and

$$\frac{\partial f}{\partial t}(x,t) = g(f(x,t),t)$$

for all $x \in M, t \in \Delta$. □

Exercise 6.11 In the proof of Proposition 6.10 we have implicitly used the fact that $[0,1]$ is closed under addition by elements from Δ. Prove it using facts established in the first chapter.

6.2 Additional Postulates of SDT

The central theme of differential topology is to reduce local to infinitesimal notions—the latter both algebraic and logical—and to exploit them in order to gain information about the former. In this section we lay down some important concepts to be employed in this context.

Definition 6.12 Let $f \in M^N$, with N and M be infinitesimally linear objects in $\mathscr{E}$. Call f a *submersion at* $x_0 \in N$ if $(df)_{x_0} : T_{x_0}N \longrightarrow T_{f(x_0)}M$ is surjective. Call f a *submersion* if f is a submersion at every $x \in N$.

Remark 6.13 The precise meaning of Definition 6.12 is that the statement

$$\forall v \in M^D\,[(\pi_0(v) = f(x_0)) \Rightarrow \exists u \in N^D(\pi_0(u) = x_0 \wedge f^D(u) = v)]$$

holds in $\mathscr{E}$, for any $f \in M^N$ and $x_0 \in N$.

Denote by $\mathrm{Subm}(M^N) \subset M^N$ the subobject of submersions. Let $f \in R^{m\Delta(n)}$. The *Jacobian* of f at $x_0 \in \Delta(n)$ is the matrix

$$D_{x_0}f = \left(\frac{\partial f_i}{\partial x_j}(x_0)\right)_{ij}.$$

Proposition 6.14 *Let $f \in R^{mR^n}$ and $x \in R^n$, both in $\mathscr{E}$. Then the following are equivalent conditions.*

(i) *f is a submersion at x.*

(ii) $\operatorname{rank}(D_x f) = m$.

(iii) $\bigvee_{(i_1,\dots,i_m)\in\binom{n}{m}} \left\{ \frac{\partial f(x)}{\partial x_{i_1}}, \dots, \frac{\partial f(x)}{\partial x_{i_m}} \right\}$ *is linearly independent.*

Exercise 6.15 Prove Proposition 6.14.

Proposition 6.16 *Let $n \geq m$. Then,* $\operatorname{Subm}(R^{m\Delta(n)}) \subset R^{m\Delta(n)}$ *is a weak open.*

Proof Since R is T_1-separated for the intrinsic topology and since $R^* = \neg\{0\}$,

$$[[A \in \mathfrak{M}_{k\times k}(R) \mid \det(A) \# 0]]$$

is an intrinsic or P-open. It follows that

$$[[A \in \mathfrak{M}_{n\times m}(R) \mid \operatorname{rank}(A) = m]] \subset R^{n\cdot m}$$

is an intrinsic or P-open and therefore a Euclidean open. Now, if the matrix A has rank m, the matrices whose entries differ from those of A by less than some $\varepsilon \in R$, $\varepsilon > 0$, will also have rank m. If f is a submersion, the rank of its differential matrix is m and therefore there exists $\varepsilon > 0$ such that $V(\Delta(n), 1, f, \varepsilon) \subset \operatorname{Subm}(R^{m\Delta(n)})$, hence the claim. □

The central theme of [96] is to state and prove a theorem of local inversion which would explain the need for Grothendieck to introduce the étale topos. Of the various versions of it, the one that we will find useful here is the following.

Postulate 6.17 (Postulate I.I) *For positive integer n,*

$$\forall f \in \Delta(n)^{\Delta(n)} \left[(f(0) = 0 \wedge \operatorname{rank}(D_0 f) = n) \Rightarrow f \in \operatorname{Iso}\left(\Delta(n)^{\Delta(n)}\right) \right].$$

The theorem below (Submersion theorem) is classically obtained as a special case of the Rank theorem [13] and, unlike the Rank theorem itself, in our context it is a consequence of the assumptions we have made so far, including Postulate I.I.

Theorem 6.18 (Submersion theorem) *Let $f \in R^{m\Delta(n)}$, $x \in \Delta(n)$, with f a submersion at x. Then $f|_x$ is locally equivalent to $\pi^n_m|_0$ where $\pi^n_m \in R^{mR^n}$ is the projection described by the rule $(x_1, \dots, x_n) \mapsto (x_1, \dots, x_m)$.*

Proof Since for any $x \in R^n$, there is an isomorphism

$$\Delta(n) \longrightarrow x + \Delta(n),$$

given by addition with x, we can restrict ourselves to the case $x = 0$ and $f(0) = 0$. Thus, instead of $f|_x$ we shall consider $f|_0$, by which we mean the composite

$$\neg\neg\{0\}^n \xrightarrow{a_x} \neg\neg x \xrightarrow{f|_x} \neg\neg\{f\} \xrightarrow{a_x{}^{-1}} \neg\neg\{0\}^m.$$

That we can do this relies on the following observations. Firstly, the local right invertibility of the Jacobian at x (which depends solely on the germ at x) is not affected by this change. Secondly, if the above composite results locally equivalent to $\pi^n_m|_0$, so will $f|_x$ itself.

We shall argue using explicitly the interpretation in $\mathscr{E}$ of the formula expressing that f is a submersion at 0 which is given in Remark 6.13. To this end, assume that f is defined at stage A in $\mathscr{E}$. In other words, both $f|_0$ and $\pi^n_m|_0$ are morphisms in the slice topos $\mathscr{E}/A$. Since f is a submersion at 0, there exists some jointly epimorphic family

$$\{\zeta_i \colon A_i \longrightarrow A\}_{i \in I}$$

in $\mathscr{E}$ such that for each $i \in I$ there is an m-tuple $(i_1, \ldots, i_m)$ such that

$$\left\{\zeta_i^*\left(\frac{\partial f(0)}{\partial x_{i_1}}\right), \ldots, \zeta_i^*\left(\frac{\partial f(0)}{\partial x_{i_m}}\right)\right\}$$

is linearly independent in $\mathscr{E}/A$.

We need to show now that for each $i \in I$ there is a jointly epimorphic family $\{\gamma_{ij} \colon B_{ij} \longrightarrow A_i\}_{j \in J_i}$ such that for each $j \in J_i$ one has $\gamma_{ij}{}^*(\zeta_i{}^* f)|_0 \sim \pi^n_m|_0$ in $\mathscr{E}/B_{ij}$. Composing coverings will give a covering $\{\zeta_k \colon B_k \longrightarrow A\}$ and the desired conclusion. The argument would therefore be the same were we to suppose that $\{\frac{\partial f(0)}{\partial x_1}, \ldots, \frac{\partial f(0)}{\partial x_m}\}$ is linearly independent. which we do now, for the sake of simplicity.

Define $\varphi \in_A R^{nR^n}$ by $\varphi = \langle f, \pi^n_{n-m} \rangle$. Clearly, $\varphi(0) = 0$ and the Jacobian of φ at 0 is given by the matrix

$$\left(\begin{array}{c|cccc} \multicolumn{5}{c}{\left(\dfrac{\partial f_j}{\partial x_i}(0)\right)_{ji}} \\ \hline & 1 & 0 & \ldots & 0 \\ * & 0 & 1 & \ldots & 0 \\ & \vdots & \vdots & \ddots & \vdots \\ & 0 & 0 & \ldots & 1 \end{array}\right) \begin{array}{l} \}\, m \\ \\ \}\, n-m \\ \\ \end{array}$$

$$\underbrace{\qquad}_{m} \qquad \underbrace{\qquad\qquad}_{n-m}$$

By assumption, $\operatorname{rank}(D_0 f) = m$. Thus, $\operatorname{rank}(D_0 \varphi) = n$ and φ is a submersion at 0. By Proposition 6.14(3), $d\varphi_0$ is locally surjective, hence bijective [59] or [61, Ex. 10.1], which from Postulate K follows that for each n, any injective linear map $R^n \longrightarrow R^n$ is bijective.

Hence, $j_0^1\varphi$ is locally an isomorphism and so, by Postulate I.I, $\varphi|_0$ is also locally an isomorphism. By the uniqueness of inverses it is enough to suppose that there is $\gamma\colon B \twoheadrightarrow A$ such that $\gamma^*(\varphi|_0) = \gamma^*\varphi|_0$ is an invertible germ in $\mathscr{E}/B$. Denote by g the composite

$$\neg\neg\{0\}^n \stackrel{\gamma^*\varphi|_0)^{-1}}{\longrightarrow} \neg\neg\{0\}^n \stackrel{\gamma^* f|_0}{\longrightarrow} \neg\neg\{0\}^m.$$

From the identity $\gamma^*\varphi|_0 \circ g = \gamma^* f|_0$ follows that if $x = (x_1,\ldots,x_n) \in \neg\neg\{0\}^n$, under $\gamma^*\varphi|_0$,

$$x \mapsto (\gamma^* f_1(x),\ldots,\gamma^* f_m(x), x_{m+1},\ldots,x_n)$$

and under g,

$$(\gamma^* f_1(x),\ldots,\gamma^* f_m(x), x_{m+1},\ldots,x_n) \mapsto (\gamma^* f_1(x),\ldots,\gamma^* f_m(x))$$

so that g is identified with $\pi^n_m|_0$. This says that $\gamma^* f|_0 \sim \pi^n_m|_0$ as required. □

In the context of SDG there are various ways to introduce the notion of manifold. One such definition that is due to Penon [94] is best suited to deal with germs of smooth mappings. The basis for it lies in the idea of an 'infinitesimal neighbourhood' of a global section $a\colon 1 \longrightarrow X$ of any object X of a given topos $\mathscr{E}$, given by

$$\neg\neg a = [[x \in X \mid \neg\neg(x = a)]].$$

With it, a notion of manifold of dimension n emerges intuitively as that of an object X for which the following holds in $\mathscr{E}$:

$$\forall x \in X[\neg\neg\{x\} = \neg\neg\{0\}^n].$$

It is proved in [94] that if M is a manifold of dimension n in the usual sense, then M regarded in a topos $\mathscr{E}$ which is a well adapted model of SDG, a notion which includes that there be an embedding $\mathscr{M}^\infty \hookrightarrow \mathscr{E}$ with certain properties, is a manifold of dimension n. It is remarked in [94] that, in general, the notion of a manifold as given therein is considerably stronger than that of Kock and Reyes [65]. The reader is referred to the indicated sources for further details about this matter.

For our purposes, however, we shall only consider a notion of 'manifold' in the following form, namely, as a 'submanifold' of R^n for some n, as follows.

Definition 6.19 Let $(\mathscr{E}, R)$ be a model of SDG. A subobject $N \subset R^n$ is said to be a *submanifold of dimension $r \leq n$ (or of codimension $r-n$)* if for each $x \in N$ there exists an isomorphism

$$\alpha \colon \neg\neg\{x\} \longrightarrow \Delta(n)$$

such that the restriction of α to $\neg\neg\{x\} \cap N$ has image $\Delta(r)$, the latter regarded as a subobject of $\Delta(n)$ by means of the rule

$$(x_1, \ldots, x_r) \mapsto (x_1, \ldots, x_r, 0, \ldots, 0).$$

One of the goals of differential topology is the classification of singularities (critical values) of germs of smooth functions in small dimensions. In our context the notions of critical and regular values of a germ are defined as usual.

Definition 6.20 Let $f \in R^{m\Delta(n)}$. We say that $y \in R^m$ is *a critical value of* f if

$$\exists x \in \Delta(n) \left(f(x) = y \wedge \bigwedge_{H \in \binom{n}{m}} \det(D_x f)_H = 0 \right)$$

holds in $\mathscr{E}$, where $\binom{n}{m}$ is the usual combinatorial object internally defined in $\mathscr{E}$. We say that $y \in R^m$ is *a regular value of* f if

$$\neg(y \in \mathrm{Crit}(f))$$

holds in $\mathscr{E}$, where $\mathrm{Crit}(f) \subset R^m$ denotes the subobject of critical values of f.

We shall investigate the role of regular values in connection with the notion of submanifold introduced earlier in Definition 6.19.

Exercise 6.21 Prove, using Postulate K, that for $y \in R^m$, y is a regular value of f if and only if the following holds in $\mathscr{E}$:

$$\forall x \in R^n \left(\neg(f(x) = y) \vee f \in \mathrm{Subm}_x(R^{mR^n}) \right).$$

Exercise 6.22 Prove that if $f \colon R^n \longrightarrow R^m$ is a submersion, then every $y \in R^m$ is a regular value of f.

Corollary 6.23 (Preimage theorem) *Let $y \in R^m$ be a regular value of $f \in R^{mR^n}$ in $\mathscr{E}$. Then $M = f^{-1}\{y\}$ is a submanifold of R^n of codimension m (i.e., dimension $n-m$).*

Proof Assume that f and x are both given at the same stage A in $\mathscr{E}$. If $x \in_A M$ then $f(x) = y$ so that f is necessarily a submersion at x. By Theorem 6.18, $f|_0$ is locally equivalent to $\pi^n_m|_0$. Thus, there is a jointly epimorphic family

$\{\gamma_i \colon B_i \longrightarrow A\}_{i \in I}$ and, for each $i \in I$, isomorphisms α_i, β_i so that

$$\begin{array}{ccc} \Delta(n) & \xrightarrow{f|_0} & \neg\neg\{f(0)\} \\ {\scriptstyle \varphi_i}\downarrow & & \downarrow{\scriptstyle \psi_i} \\ \Delta(n) & \xrightarrow[\pi^n_m|_0]{} & \Delta(m) \end{array}$$

commutes in $\mathscr{E}/B_i$. (We do not change the notation of the projections when passing from $\mathscr{E}$ to $\mathscr{E}/A$ or from $\mathscr{E}/A$ to $\mathscr{E}/B_i$ since these functors preserve products. We have omitted the notation indicating the change of stage.) The result now follows from the commutativity of the diagram above by virtue of the following chain of isomorphisms:

$$(f|_x)^{-1}\{\psi_i(0)\} \cong \neg\neg\{x\} \cap M \cong (\pi^n_m)^{-1}\{0\} \cong \Delta(n-m)\,.$$

□

Remark 6.24 The result of Corollary 6.23 establishes in our context that the solutions of an equation $y = f(x)$ form a submanifold of R^n provided $y \in R^m$ is a regular value of $f \in R^{mR^n}$. In order to extend a result of this sort to subobjects $N \subset R^n$ and not just elements of R^n (or of $\Delta(n)$ if we consider germs) we need to impose a suitable condition on N that generalizes that of a regular element. This is where the concept of transversality comes in. The category $\mathscr{M}^\infty$ of C^∞-manifolds and C^∞-mappings does not have all inverse limits. It is well known that if $f \colon M \longrightarrow \mathbb{R}$ is a differentiable mapping, then $f^{-1}(\{0\})$ is, in general, just an arbitrary closed subset of M. Moreover, for a C^∞-mapping $f \colon M \longrightarrow N$ between manifolds, and $U \subset N$ a submanifold, in order for $f^{-1}(U) \subset M$ to be a submanifold, f and U must be 'well positioned'. This notion can be made precise in differential topology through that of transversality. We introduce next a notion of transversality in our context which, just as in the classical theory, is a generalization of the notion of a regular value [51].

Definition 6.25 Let $f \in R^{m\Delta(n)}$ and let $N \subset R^m$ be any subobject. Let $x \in \Delta(n)$ such that $f(x) \in N$. We say that *f is transversal to N at x*, and write $f \pitchfork_x N$, if

$$T_{f(x)}R^m = \mathrm{Im}(df)_x + T_{f(x)}N\,.$$

We say that *f is transversal to N* if

$$\forall x \in \Delta(n)\,(f \pitchfork_x N)\,.$$

Below are some graphic examples.

It will be convenient to avail ourselves of an additional postulate of density

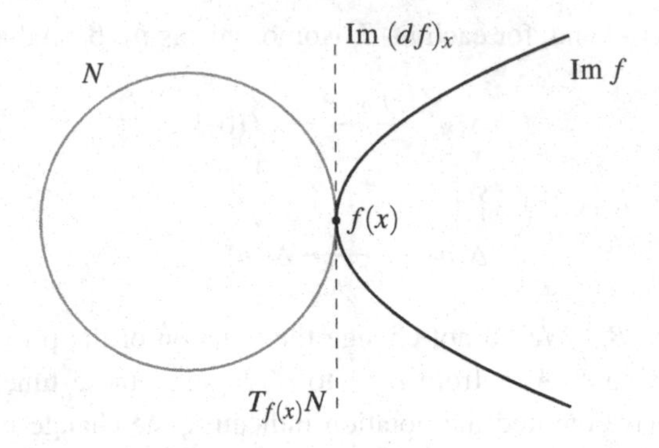

Figure 6.1 f not transversal to N

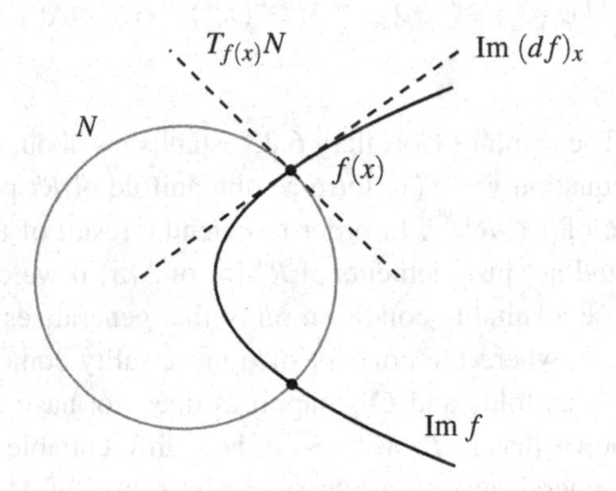

Figure 6.2 f transversal to N

of regular values. This postulate is consistent with Sard's theorem [104] in classical differential topology, which states that the set of critical values for any smooth mapping $f\colon \mathbb{R}^n \longrightarrow \mathbb{R}^m$ has measure 0.

Postulate 6.26 (Postulate D)

$$\forall U \in P(R^m)\, \forall f \in R^{m\Delta(n)}\, \exists y \in U\, \neg(y \in \mathrm{Crit}(f))\,.$$

We are now in a position to list the axioms and postulates of the theory we call SDT.

Definition 6.27 By *synthetic differential topology* (**SDT**) we shall mean the data consisting of a ringed topos $(\mathscr{E}, R)$ satisfying the axioms and postulates of SDG which in addition satisfies the following axioms and postulates :

- Axiom **G** (germs representability),
- Axiom **M** (the germ representing objects are tiny),
- Postulate **E** (the Euclidean topological structure satisfies the covering principle),
- Postulate **S** (existence and uniqueness of solutions to parametrized ordinary differential equations),
- Postulate **I.I** (infinitesimal inversion), and
- Postulate **D** (density of regular values).

In the rest of this section we assume that $(\mathscr{E}, R)$ is a model of SDT.

Definition 6.28 A germ $f \in R^{m\Delta(n)}$ is said to be an *immersion* if $\mathrm{Rank}(D_0 f) = n$.

Proposition 6.29 *If $m > 2n$ then the class of immersions is dense in R^{mR^n} for the weak topological structure.*

Proof Our aim is to show that any basic neighbourhood of the weak topology contains an immersion. Recall that since $\Delta(n)$ is compact, we may restrict our considerations to basic W-open objects of the form $V(\Delta(n), r, h, \varepsilon)$ with $h \in R^{mR^n}$, $1 \le r \le n$ and $\varepsilon \in R$, $\varepsilon > 0$. For any given such an object we will show that there exists a polynomial $\sigma \in R^{mR^n}$ of total degree l and coefficients $c_i \in (-\varepsilon, \varepsilon)^m$ such that $h + \sigma|_{\Delta(n)}$ is an immersion.

Let $s = \mathrm{Rank}(D_0 h)$ and define $\Phi \in R^{m\Delta(s+n)}$ as follows:

$$\Phi(\lambda, x) = \sum_{i=1}^{s} \lambda_i \frac{\partial h}{\partial x_i}(x) - \frac{\partial h}{\partial x_{s+l}}(x).$$

By Postulate D, we have

$$\exists c_{s+l} \in R^m \left(c_{s+l} \in (-\varepsilon, \varepsilon)^m \wedge c_{s+l} \in \mathrm{Reg}(\Phi)\right)$$

where $\mathrm{Reg}(\Phi) \subset R^m$ is the subobject of regular values of Φ.

Define $g_l \in R^{mR^n}$ by $g_l(x) = h(x) + c_{s+l} \cdot x_{s+l}$. By ordinary differentiation we get

$$\frac{\partial g_l}{\partial x_i}(x) = \frac{\partial l_i}{\partial x_i}(x)$$

for every $x \in \Delta(n)$, $i \le s$, and

$$\frac{\partial g_l}{\partial x_{s+l}}(x) = \frac{\partial l_i}{\partial x_{s+l}}(x) + c_{s+l}$$

for every $x \in \Delta(n)$.

Since c_{s+l} is a regular value of Φ and $s \leq n$, $p \geq 2n$, Φ cannot be a submersion at (λ, x) in the sense that

$$\forall(\lambda,x) \in \Delta(s+n) \neg(\Phi \in \mathrm{Subm}_{(\lambda,x)}(R^{m\Delta(s+n)})),$$

and c_{s+l}, being a regular value of Φ, cannot be in the image of Φ. In particular, in the internal sense that this is the case, we must have $\neg(\Phi(0,0) = c_{s+l})$. Using this remark and the fact that $s = \mathrm{rank}(D_0 h)$, it is easily seen that

$$\forall\lambda_1,\ldots,\lambda_s \in R \neg\Big(\sum_{i=1}^{s} \lambda_i \frac{\partial l_i}{\partial x_i}(0) - \frac{\partial l_i}{\partial x_{s+l}}(0) = c_{s+l}\Big)$$

which means that

$$\forall\lambda_1,\ldots,\lambda_s \in R \neg\Big(\sum_{i=1}^{s} \lambda_i \frac{\partial g_l}{\partial x_i}(0) = \frac{\partial g_l}{\partial x_{s+l}}(0)\Big)$$

and this amounts to stating that

$$\Big\{\frac{\partial g_1}{\partial x_1}(0),\ldots,\frac{\partial g_l}{\partial x_s}(0),\frac{\partial g_l}{\partial x_{s+l}}(0)\Big\}$$

is linearly independent. By repeating this procedure $n-(s+l)$ times we get, successively, elements $c_{s+1},\ldots,c_n \in (-\varepsilon,\varepsilon)^m$, coefficients of the desired polynomial $\sigma(x) = c_{s+1}\cdot x_{s+1} + \cdots + c_n \cdot x_n$, as $h+\sigma$ is an immersion, and certainly $h+\sigma \in V(\Delta(n), r, h, \varepsilon)$. □

We mention two more results about immersions that can be shown in this context.

Proposition 6.30 *If $m \geq n$, the object* $\mathrm{Imm}(R^{mR^n}) \subset R^{mR^n}$ *is a weak open.*

Proposition 6.31 *If $f \in R^{mR^n}$ is an immersion at $0 \in R^n$, then f is infinitesimally injective at* 0.

Exercise 6.32 Prove Proposition 6.30 and Proposition 6.31.

To end this section we give, within SDT, a proof of Thom's Transversality Theorem [47], which a key result in the classical theory of stable mappings and their singularities.

Definition 6.33 (i) We say that $g_1,\ldots,g_s \in R^{R^n}$ are *independent functions* if

$$\langle g_1,\ldots,g_s\rangle : R^n \longrightarrow R^s$$

is a submersion at any $x \in \bigcap_{i=1}^{s} g_i^{-1}(0)$.

(ii) A submanifold $M \subset R^n$ is said to be *cut out by independent functions* if

$$M = \bigcap_{i=1}^{s} g_i^{-1}(0)$$

where $g_1, \ldots, g_s$ are independent functions.

Theorem 6.34 *Let $f \in R^{mR^n}$ and $N \subset R^m$ be a submanifold cut out by independent functions and of codimension $s \leq n$. If $f \pitchfork N$, then $M = f^{-1}(N) \subset R^n$ is a submanifold of dimension s, also cut out by independent functions.*

Remark 6.35 In the classical setting [51], the notion of manifold is defined by means of local concepts and it is the case that every submanifold is locally cut out by independent functions. For our notion of submanifold, which resorts to infinitesimal rather than local notions, the same result can be obtained on account of Axiom G and Postulate I.I., as in the next proposition.

Proposition 6.36 *Every submanifold $N \subset R^n$ is cut out by independent functions, that is, for each $x \in N$, there exist independent functions $g_1, \ldots, g_s$ such that*

$$\neg\neg\{x\} \cap N = \bigcap_{i=1}^{s} \{g_i{}^{-1}(0)\}.$$

Proof Let $N \subset R^n$ be a submanifold of dimension r. For any $x \in N$ there exists an isomorphism

$$\alpha : \neg\neg\{x\} \longrightarrow \Delta(n)$$

whose restriction to $\neg\neg\{x\} \cap N$ has its image in $\Delta(r)$, the latter identified with $\Delta(r) \times \{0\} \subset \Delta(n)$. Hence, taking $g_i = \pi_{r+i} \circ \alpha$, for $i = 1, \ldots, n-r$, given that the projections are submersions and α is a diffeomorphism, the claim follows. □

The next result exhibits transversality as a submersion condition.

Proposition 6.37 *Let $f \in R^{mR^n}$ and let $N \subset R^m$ be a submanifold cut out by independent functions $g_1, \ldots, g_s$. Then, the following are equivalent conditions.*

(i) $f \pitchfork_x N$.

(ii) *The composite $\langle g_1, \ldots, g_s \rangle \circ f$ is a submersion at x.*

Proof Let $g = \langle g_1, \ldots, g_s \rangle$. It is easy to verify that

$$\text{Ker}\,(dg)_{f(x)} = T_{f(x)}N$$

and therefore

$$T_{f(x)}R^m = \mathrm{Im}\,(df)_x + T_{f(x)}N = \mathrm{Im}\,(df)_x + \mathrm{Ker}\,(dg)_{f(x)}\,.$$

Since g is a submersion at x, $g \circ f$ is a submersion at x if and only if

$$T_{f(x)}R^m = \mathrm{Im}\,(df)_x + \mathrm{Ker}\,(dg)_{f(x)}$$

which is equivalent to $f \pitchfork_x N$. □

The following constitutes a generalization of the preimage theorem (Corollary 6.23).

Theorem 6.38 *Let $f \in R^{mR^n}$ and $N \subset R^m$ a submanifold of codimension $s \leq m$ cut out by independent functions. Assume that $f \pitchfork N$. Then, $M = f^{-1}(N) \subset R^n$ is a submanifold of codimension s (also cut out by independent functions).*

Proof Let $f \in R^{mR^n}$ and $N \subset R^m$ be given at stage A, and assume that $f \pitchfork N$. By definition of submanifold cut out by independent functions, there is a jointly epimorphic family $\{A_i \twoheadrightarrow A\}_{i \in I}$ so that, for each $i \in I$, N is carved out of R^m by independent functions $g^i_1, \ldots, g^i_s \in_{A_i} R^{R^n}$. Define a new function $g^i = \langle g^i_1, \ldots, g^i_s \rangle \in_{A_i} R^{sR^n}$. We claim that $g^i \circ f$ is a submersion at every $x \in_{A_i} R^n$ for which $g^i \circ f(x) \in N$. To see this, use the following commutative diagram in $\mathscr{E}/A_i$:

$$\begin{array}{ccc}
T_xR^n & \xrightarrow{\;d(g^i f)_x\;} & T_{g^i f(x)}R^s \\
{\scriptstyle df_x}\big\downarrow & & \big\uparrow{\scriptstyle (dg^i)_{f(x)}} \\
\mathrm{Im}\,df_x & \xrightarrow[\;\pi^n_m|_0\;]{} & T_{f(x)}R^m
\end{array}$$

Since g^i is a submersion, $(dg)_{f(x)}$ is locally surjective and the result follows from the definition of g^i and the condition

$$T_{f(x)}R^m = \mathrm{Im}\,df_x + \mathrm{Ker}\,(dg^i)_{f(x)} = \mathrm{Im}\,df_x + T_{f(x)}N$$

at stage A_i. The second equality follows from definition of g^i and so

$$T_{f(x)}R^m = \mathrm{Im}\,df_x + T_{f(x)}N$$

which is precisely what transversality says at stage A_i. Using now Corollary 6.23, $(g^i \circ f)^{-1}\{0\}$ is a submanifold of codimension s, and we have the equalities

$$(g^i \circ f)^{-1}\{0\} = f^{-1}(g^{-1}\{0\}) = f^{-1}(N)$$

which ends the proof. □

Theorem 6.39 (Thom's Transversality Theorem) *For $n, m > 0$ and $1 \leq r \leq n$, given any $N \subset R^{mD_r(n)} = R^s$ a submanifold cut out by independent functions, the class of germs $g \in R^{m\Delta(n)}$ with $J^r g \pitchfork N$ is dense for the weak topological structure.*

Proof With the same simplifications as in Proposition 6.29, given a basic W-open object of the form $V(\Delta(n), r, h, \varepsilon)$ with $h \in R^{mR^n}$, $1 \leq r \leq n$ and $\varepsilon \in R$, $\varepsilon > 0$, we will find a polynomial $\sigma \in R^{mR^n}$ of total degree l and coefficients $c_i \in (-\varepsilon, \varepsilon)^m$ such that $J^r(h + \sigma|_{\Delta(n)}) \pitchfork N$. Define γ_h at level A, given by the following rule:

$$[(x, f) \in \Delta(n) \times R^{mD_r(n)} \mapsto J^r(h + f)(x) \in R^{mD_r(n)}].$$

It follows easily from the identification $R^{mD_r(n)} \cong R^s$ in investigating the Jacobian of γ_h, that the latter is a submersion and therefore $\gamma_h \pitchfork N$. Since N is cut out by independent functions, Corollary 6.23 gives that $M = \gamma_h{}^{-1}(N)$ is a submanifold of $\Delta(n) \times R^s$ and so by Postulate D we get

$$\exists (c_{i,\alpha})_{1 \leq i \leq m,\, 1 \leq \alpha \leq \binom{n+k}{k}} \in R^s\, [c_{i,\alpha} \in (-\varepsilon, \varepsilon) \wedge (c_{i,\alpha}) \in \mathrm{Reg}(\pi^M|_0)],$$

where $\pi^M|_0$ denotes the germ at 0 of the restriction to M of the projection map $\pi\colon \Delta(n) \times R^s \longrightarrow R^s$.

Define $\sigma_i(x) = \sum_{|\alpha| \leq t} c_{i,\alpha} \cdot x^\alpha$ for $i = 1, \ldots, m$ and check that $\sigma = (\sigma_i)_{1 \leq i \leq m}$ is the required polynomial. □

PART FOUR

TOPICS IN SDT

In this fourth part we present, within the context of synthetic differential topology (SDT), a theory of stable germs of smooth maps including Mather's theorem on the equivalence between stability and infinitesimal stability, followed by Morse theory. Germ representability by logico-infinitesimal objects brings about a considerable simplification of the subject.

7

Stable Mappings and Mather's Theorem in SDT

In this chapter we present, within the context of SDT, the preliminaries to a theory of stable germs of smooth mappings and their singularities. The notion of stability for mappings, germs, or unfoldings is important for several reasons, one of which being its intended application in the natural sciences, and as promoted by R. Thom. Another reason for concentrating on stability is in the simplification that it brings to the classification of singularities. A first proof of Mather's theorem ('stability if and only if infinitesimal stability') was given in [44, 26]. In order to prove the hard part of the theorem ('infinitesimal stability implies stability') it was resorted therein to the Malgrange preparation theorem in geometric form and assumed as a postulate, which was then showed to be valid in a topos model of SDT. In [103], a second proof of Mather's theorem was given that does not use the Malgrange preparation theorem, but which instead introduces a new notion of (transversal) stability for this purpose. The theory of stability of germs of smooth functions can thus be developed using only the given axioms of SDT. We then give an application of Mather's theorem to the theory of Morse germs within SDT.

7.1 Stable Mappings in SDT

Among all physical quantities (in the sense that they are part of some theory in physics) it is natural to single out those which remain 'the same' when slightly deformed. Such quantities are said to be 'stable'.

For instance, a germ (at 0) of a smooth function $f\colon \mathbb{R}^n \longrightarrow \mathbb{R}^m$ (which we may assume is such that $f(0) = 0$) is 'stable' provided any other such germ of a function $g\colon \mathbb{R}^n \longrightarrow \mathbb{R}^m$ which is 'near' f is equivalent to it, that is, the germ of g can be deformed by means of germs of diffeomorphisms φ on $\mathbb{R}^n$ and ψ on $\mathbb{R}^m$ to become f, in the sense that $g = \psi \circ f \circ \varphi^{-1}$. We illustrate the idea of equivalence for functions (or their germs) by means of the figures below. In Fig. 7.1, the function represented by a thick line is equivalent to that drawn by a thin line, whereas in Fig. 7.2 the functions so described are not equivalent although they are 'near to each other'.

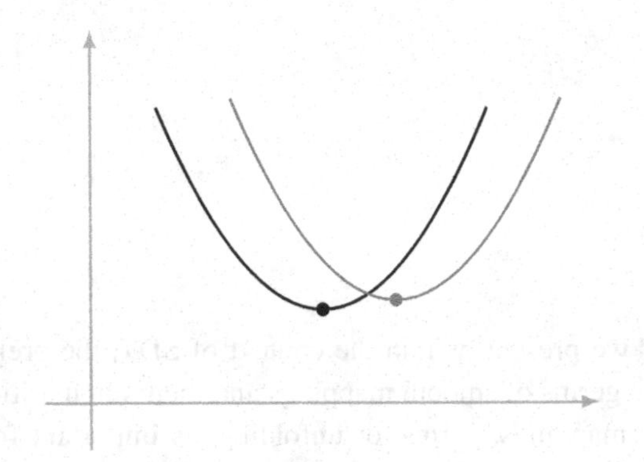

Figure 7.1 Equivalent maps

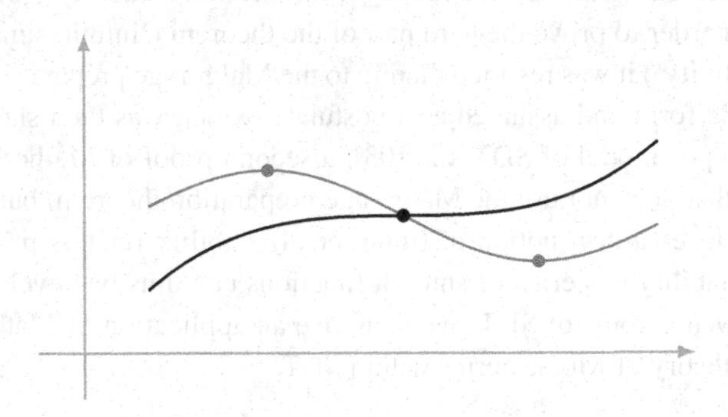

Figure 7.2 Non-equivalent maps

These notions and statements were introduced in [20] employing the Penon opens [96] to express the notion of 'nearness' which is basic to stability. By Axiom G and Axiom M of a topos model $(\mathscr{E}, R)$ of SDT, germs are represented by a tiny object, hence can be manipulated directly instead of having to take representatives of equivalence classes. This is one of the advantages of our treatment.

The basic ingredients in a (classical) theory of stability for C^∞-mappings are a notion of 'nearness' and a notion of 'equivalence' and similarly for germs.

In what follows, $(\mathscr{E}, R)$ is assumed to be a model of SDT as in Definition 6.27. For germs $f, g \in R^{m\Delta(n)}$ of smooth mappings in SDT (by virtue of Axiom G), we need to express the idea that f is *equivalent* to g if they are 'the same' up to isomorphisms on the domain and the range. This can be made precise after establishing some terminology.

Definition 7.1 We say that $\varphi \in R^{nR^n}$ is *infinitesimally invertible* (respectively *infinitesimally surjective*) *at* $x \in R^n$ if the restriction

$$\varphi|_x\colon \neg\neg\{x\} \longrightarrow \neg\neg\{\varphi(x)\}$$

is an isomorphism (respectively a surjection.)

Definition 7.2 Let $f, g \in R^{m\Delta(n)}$. Assume that $f(0) = y$ and $g(0) = v$. Say that $f \sim g$ if the following holds:

$$\exists x, u \in R^n\ \exists \varphi \in R^{nR^n}\ \exists \psi \in R^{mR^m}\ \big[\ \varphi \text{ inf. inv. at } x \wedge \psi \text{ inf. inv. at } y \\ \wedge\ \varphi(x) = u \wedge g_u = (\psi|_y) \circ f_x \circ (\varphi|_x)^{-1}\big],$$

where f_x is the composite

$$\neg\neg\{x\} \xrightarrow{a_x^{-1}} \Delta(n) \xrightarrow{f} \neg\neg\{y\}$$

and a_x is the isomorphism 'addition with x', and similarly for g_u.

In a diagram, the condition is stated as the commutativity of the following diagram where $v = \psi(y)$.

$$\begin{array}{ccc} \neg\neg\{x\} & \xrightarrow{f_x} & \neg\neg\{y\} \\ {\scriptstyle \varphi|_x}\downarrow & & \downarrow{\scriptstyle \psi|_y} \\ \neg\neg\{u\} & \xrightarrow[g_u]{} & \neg\neg\{v\} \end{array}$$

Remark 7.3 We shall from now on consider only germs at 0 taking 0 to 0, as this is an inessential restriction which renders the notations and calculations considerably simpler. The nearness condition on function spaces of the form $R^{m\Delta(n)}$, that is, on germs at 0 of smooth mappings from R^n to R^m, will be obtained from the *weak topological structure*, introduced in Chapter 5 for function spaces R^{nX}.

Denote by

$$G = \mathrm{Inv}\big(\Delta(n)^{\Delta(n)}\big) \times \mathrm{InfInv}_0\big(R^{mR^m}\big)$$

where

$$\mathrm{Inv}\left(\Delta(n)^{\Delta(n)}\right) = [[\varphi \in \Delta(n)^{\Delta(n)} \mid \varphi \text{ invertible}]]$$

and

$$\mathrm{Inf\,Inv}_0(R^{mR^m}) = [[\psi \in R^{mR^m} \mid \psi \text{ inf. inv. at } 0]]\,.$$

Define

$$\gamma_f \colon G \longrightarrow R^{m\Delta(n)}$$

by the formula $\gamma(\varphi,\psi) = \psi \circ f \circ \varphi^{-1}$. Notice that

$$\gamma_f(\mathrm{id}_n, \mathrm{id}_m) = f$$

where id_n denotes the identity endomorphism of $\Delta(n)$ and id_m denotes the identity endomorphism of R^m. It follows from Definition 7.2 that for any germ $f \in R^{m\Delta(n)}$, and $(\varphi,\psi) \in G$, $\gamma_f(\varphi,\psi) \sim f$.

Definition 7.4 A germ $f \in R^{m\Delta(n)}$ is said to be *stable* if $\mathrm{Im}\,(\gamma_f)$ is a weak open of $R^{m\Delta(n)}$. It is also assumed that if $\gamma_f(\varphi,\psi) = f$ then $\varphi = \mathrm{id}_n$ and $\psi = \mathrm{id}_m$.

Recall from Chapter 2 that for an infinitesimally linear object M and $x \in M$, the object T_xM of tangent vectors at x is an R-module and that any map between infinitesimally linear, $h\colon M \longrightarrow N$ and $x \in M$, induces a linear map

$$(dh)_x \colon T_xM \longrightarrow T_{h(x)}N\,,$$

called the differential of h at x.

Definition 7.5 A germ $f \in R^{m\Delta(n)}$ is said to be *infinitesimally stable* if γ_f is a submersion at $(\mathrm{id}_n, \mathrm{id}_m)$, that is to say that

$$(d\gamma_f)_{(\mathrm{id}_n,\mathrm{id}_m)} \colon T_{(\mathrm{id}_n,\mathrm{id}_m)}G \longrightarrow T_f\left(R^{m\Delta(n)}\right)$$

is surjective.

We can now give a proof within SDT of the easy part of Mather's theorem which is in essence a theorem of local inversion.

Theorem 7.6 *(Stability implies infinitesimal stability) For any $f \in R^{m\Delta(n)}$ (with $f(0) = 0$), if f is stable, then f is infinitesimally stable.*

Proof Since f is stable, and since the weak topological structure is subintrinsic, it follows that

$$\forall g \in R^{m\Delta(n)}\left(\neg(g = f) \vee g \in \mathrm{Im}\,\gamma_f\right),$$

and by pure logical reasons, $\gamma_f|_{(\mathrm{id}_n, \mathrm{id}_m)}: \neg\neg(\mathrm{id}_n, \mathrm{id}_m) \longrightarrow \neg\neg(f)$ is surjective. By Axiom W for $W = D$, which says in particular that D is an atom, it follows that

$$(\gamma_f)^D : (\neg\neg\{(\mathrm{id}_n, \mathrm{id}_m)\})^D \longrightarrow (\neg\neg\{f\})^D$$

is surjective.

We argue next that the above implies that

$$T_{(\mathrm{id}_n, \mathrm{id}_m)}(\neg\neg\{(\mathrm{id}_n, \mathrm{id}_m)\}) \longrightarrow T_f(\neg\neg\{f\})$$

is surjective.

Consider $\xi \in (\neg\neg\{f\})^D$. Then, by surjectivity of $(\gamma_f)^D$, there exists $\zeta \in (\neg\neg\{(\mathrm{id}_n\,, \mathrm{id}_m)\})^D$ such that for any $d \in D$, $\xi(d) = \gamma_f(\zeta(d))$, and if $\xi(0) = f$ then by the second part of the condition on stability of f we have that $\zeta(0) = (\mathrm{id}_n\,, \mathrm{id}_m)$, that is to say, $\zeta \in T_{(\mathrm{id}_n, \mathrm{id}_m)}(\neg\neg\{(\mathrm{id}_n\,, \mathrm{id}_m)\})$. This finishes the proof since the latter agrees with

$$(d\gamma_f)_{(\mathrm{id}_n, \mathrm{id}_m)} : T_{(\mathrm{id}_n, \mathrm{id}_m)}G \longrightarrow T_f(R^{m\Delta(n)})\,.$$

$\square$

In the rest of this section we analyse the morphism $(d\gamma_f)$ so as to apply it to establish stability and non-stability properties in some examples.

Recall from section 2.3 that given any X, the object of vector fields on X, $\mathrm{Vect}(X)$, can be identified with the object $T_{\mathrm{id}_X}X^X$ of infinitesimal deformations of the identity map. In a similar way, $\mathrm{Vect}(f)$, the object of vector fields along $f \in R^{m\Delta(n)}$, that is to say, maps $\xi : \Delta(n) \longrightarrow R^{mD}$ such that $\pi \circ \xi = f$, can be identified with $T_f(R^{m\Delta(n)})$, infinitesimal deformations of f, as shown in Fig. 7.3, where $\psi : D \longrightarrow R^{m\Delta(n)}$ is the map in $\mathscr{E}$ induced by ξ.

We will see that f infinitesimally stable means precisely that any infinitesimal deformation of f is equivalent to f.

Notice that G is infinitesimally linear (see Prop. 2.3.8) and that $T_{(\mathrm{id}_n, \mathrm{id}_m)}G$ can be identified with $T_{\mathrm{id}_n}(\Delta(n)^{\Delta(n)}) \times T_{\mathrm{id}_m}(R^{mR^m})$, as $\mathrm{Inv}\,(\Delta(n)^{\Delta(n)}) \subset \Delta(n)^{\Delta(n)}$ and $\mathrm{InfInv}_0\,(R^{mR^m}) \subset R^{mR^m}$ are both Penon opens. Hence $T_{id_n}(\mathrm{Inv}\,(\Delta(n)^{\Delta(n)}) = T_{id_n}\,(\Delta(n)^{\Delta(n)})$ and therefore $T_{id_m}(\mathrm{InfInv}_0\,(R^{mR^m})) = T_{id_m}\,(R^{mR^m})\,.$

Again, any $\sigma \in T_{\mathrm{id}_n}(\Delta(n)^{\Delta(n)})$ may be regarded as an infinitesimal deformation of id_n or as a vector field on $\Delta(n)$. It follows from the definition of a vector field that for each $d \in D$, $\sigma(d) \in \Delta(n)^{\Delta(n)}$ has $\sigma(\ d)$ as its inverse.

Consider the morphism

$$\alpha_f : T_{\mathrm{id}_n}(\Delta(n)^{\Delta(n)}) \longrightarrow T_f(R^{m\Delta(n)})$$

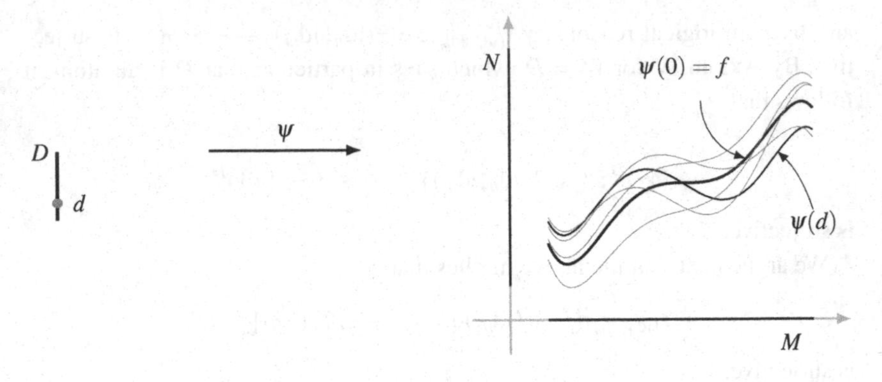

Figure 7.3 Infinitesimal deformations of $f \in N^M$

to be induced by the rule

$$\varphi \in \mathrm{Inv}\left(\Delta(n)^{\Delta(n)}\right) \mapsto f \circ \varphi^{-1} \in R^{m\Delta(n)},$$

and given by $\alpha_f(\sigma)(d) = f \circ (\sigma(d))^{-1}$ where $\sigma \in \left(\Delta(n)^{\Delta(n)}\right)^D$ is so that $\sigma(0) = \mathrm{id}_n$.

Similarly, one can define

$$\beta_f \colon T_{\mathrm{id_m}}\left(R^{mR^m}\right) \longrightarrow T_f\left(R^{m\Delta(n)}\right)$$

by $\beta_f(\tau)(d) = \tau(d) \circ f$, where $\tau \in (R^{mR^m})^D$ is so that $\tau(0) = \mathrm{id}_m$, induced by

$$\psi \in R^{mR^m} \mapsto \psi \circ f \in R^{m\Delta(n)}$$

and in the vector fields version, $\beta_f(\tau)(x) = \tau(f(x))$.

Fig. 7.4 shows the actions of α_f and β_f on the tangent vectors $\sigma(x) \in T_x\Delta(n)$ and $\tau(y) \in T_yR^m$, while Fig. 7.5 depicts the action of α_f and β_f on the corresponding germs $\sigma(d) \in \Delta(n)^{\Delta(n)}$ and $\tau(d) \in R^{m\Delta(n)}$.

Lemma 7.7

$$(d\gamma_f)_{(\mathrm{id}_n,\mathrm{id}_m)}(\sigma,\tau) = \alpha_f(\sigma) + \beta_f(\tau).$$

Proof Recall from Corollary 2.30 that $\alpha_f(\sigma) + \beta_f(\tau)$ is the tangent vector whose action on $d \in D$ is defined by $\ell(d,d)$, where $\ell \colon D(2) \longrightarrow R^{m\Delta(n)}$ is the unique map given by infinitesimal linearity of $R^{m\Delta(n)}$ such that $\ell \circ \mathrm{inc}_1 = \alpha_f(\sigma)$ and $\ell \circ \mathrm{inc}_2 = \beta_f(\tau)$. Define $\ell(d_1,d_2) = \tau(d_2) \circ f \circ (\sigma(d_1))^{-1}$ and use that $\sigma(0) = \mathrm{id}_n$ and $\tau(0) = \mathrm{id}_m$. The assertion now follows by the universal

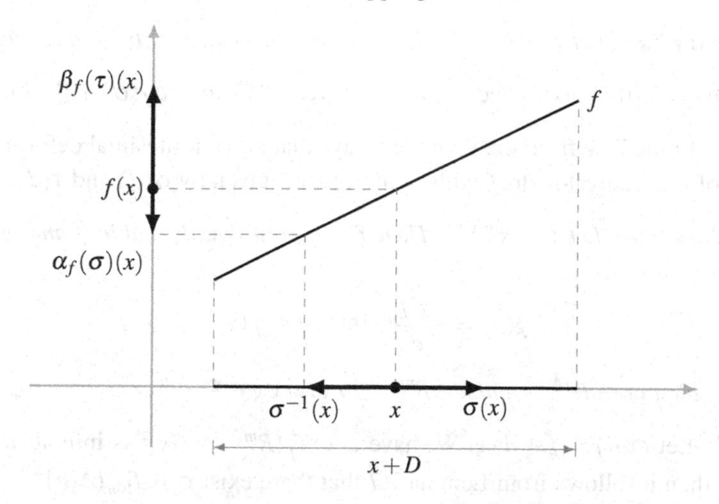

Figure 7.4 Action of α_f and β_f on vector fields

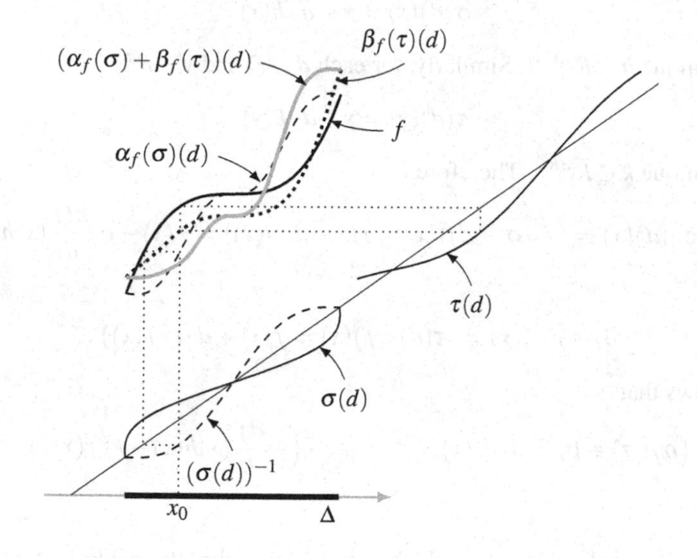

Figure 7.5 α_f and β_f on tangent vectors on spaces of germs

cancellation of the $d \in D$, as we have, for any $d \in D$,

$$(d\gamma_f)_{(\mathrm{id}_n,\mathrm{id}_m)}(\sigma,\tau)(d) = \tau(d) \circ f \circ \sigma(-d) = (\alpha_f(\sigma)+\beta_f(\tau))(d)\,.$$

□

Corollary 7.8 *Let $f \in R^{m\Delta(n)}$. Then f is infinitesimally stable if and only if*

$$\forall \omega \in \mathrm{Vect}(f)\, \exists \sigma \in \mathrm{Vect}(\Delta(n))\, \exists \tau \in \mathrm{Vect}(R^m)\, [\omega = \alpha_f(\sigma) + \beta_f(\tau)]\,.$$

Proof Immediate from the Lemma, it says that any infinitesimal deformation $\omega(d)$ of f is equivalent to f with isomorphisms given by $\sigma(d)$ and $\tau(d)$. □

Corollary 7.9 *Let $f \in R^{m\Delta(n)}$. Then f is infinitesimally stable if and only if the equation*

$$g(x) = -\frac{df}{dx}(x)h(x) + k(f(x))$$

is solvable in $h \in R^{n\Delta(n)}$ and $k \in R^{mR^m}$ for every $g \in R^{m\Delta(n)}$.

Proof Let $\omega(d) = f + d \cdot g$. We have $\omega \in T_f(R^{m\Delta(n)})$. If f is infinitesimally stable then it follows from Lemma 7.7 that there exist $\sigma \in T_{\mathrm{id}_n}(\Delta(n)^{\Delta(n)})$ and $\tau \in T_{\mathrm{id}_m}(R^{mR^m})$ such that $\alpha_f(\sigma) + \beta_f(\tau) = \omega$. Now, for each $d \in D$ and $x \in \Delta(n)$,

$$\sigma(d)(x) = x + d \cdot h(x)$$

for a unique $h \in R^{n\Delta(n)}$. Similarly, for each $d \in D$ and $y \in R^m$,

$$\tau(d)(y) = y + d \cdot k(y)$$

for a unique $k \in R^{mR^m}$. Therefore,

$$\alpha_f(\sigma)(d)(x) = (f \circ \sigma(-d))(x) = f(x - d \cdot h(x)) = f(x) - d \cdot \frac{df}{dx}(x)h(x)$$

and

$$\beta_f(\tau)(d)(x) = (\tau(d) \circ f)(x) = f(x) + d \cdot k(f(x))\,.$$

It follows that

$$\big(\alpha_f(\sigma) + \beta_f(\tau)\big)(d)(x) = f(x) + d \cdot \Big(-\frac{df}{dx}(x)h(x) + k(f(x))\Big)\,.$$

□

As an example of a germ which is not infinitesimally stable is $f(x) = x^3$. Indeed, let ω be a vector field along f with non-constant principal part on D, for instance $\omega(d)(x) = x^3 + d \cdot x$. Since for any $\sigma \in T_{\mathrm{id}}(\Delta^\Delta)$ and $d \in D$, one has $\alpha_f(\sigma)(d) \in \Delta^\Delta$, and since $f'(d) = 0$, the restriction $\alpha_f(\sigma)(d)|_D = 0$. Similarly one can argue that for any $\tau \in T_{\mathrm{id}}(R^R)$ and $d \in D$, $\beta_f(\tau)(d) \in R^R$. In this case, since f is constant on D, the restriction $\beta_f(\tau)(d)|_D$ has a constant principal part. Therefore so does $(\alpha_f(\sigma) + \beta_f(\tau))(d)$ and thus it cannot agree with $\omega(d)$ on D.

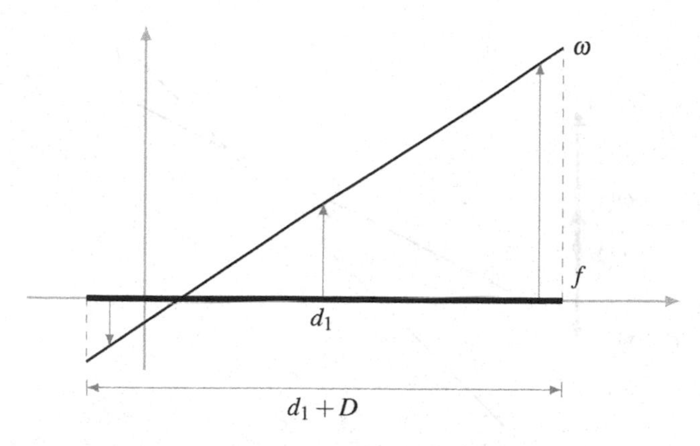

Figure 7.6 A vector field ω along $f(x)=x^3$

The argument just given to show that the function $f(x)=x^3$ is not infinitesimally stable (hence not stable) can be carried out for any $f \in R^{mR^n}$ that is constant on a subobject of R^n containing $D_2(n)$.

An instance of an infinitesimally stable germ is $f(x)=x^2$. Indeed, let ω be a vector field along f defined by means of $\omega(d)=f+d\cdot l$, where $l \in R^{\Delta}$. Using Postulate F there exists $g \in R^{R^2}$ such that for each $x \in \Delta$,

$$l(x)-l(0)=g(x,0)(x-0)=g(x,0)\,x\,.$$

It will therefore be enough to let $\sigma \in T_{\mathrm{id}}(\Delta^{\Delta})$ and $\tau \in T_{\mathrm{id}}(R^R)$ with principal parts $h(x)=-\frac{1}{2}g(x,0)$ and $k(x)=l(0)$, respectively. Thus, $\omega=\alpha_f(\sigma)+\beta_f(\tau)$ as required.

7.2 Mather's Theorem in SDT

We shall deal with a proof of the implication 'infinitesimally stable implies stable', that part of Mather's theorem that, together with Thom's transversality theorem, is important in the classification of singularities with respect to equivalence. Our proof in this section follows rather closely that of [97], but differs from it in that we deal directly with germs as if they were mappings.

The passage from the infinitesimal to the local in the proof presented here is done in two steps. From the infinitesimal stability of a germ $f \in R^{m\Delta(n)}$ — interpreted as the condition that the differential at $(\mathrm{id_n},\mathrm{id_m}) \in G$ of the associated morphism $\gamma_f\colon G \longrightarrow R^{m\Delta(n)}$ is surjective—a similar condition is postulated to hold for any germ in some neighbourhood of f in the weak topological

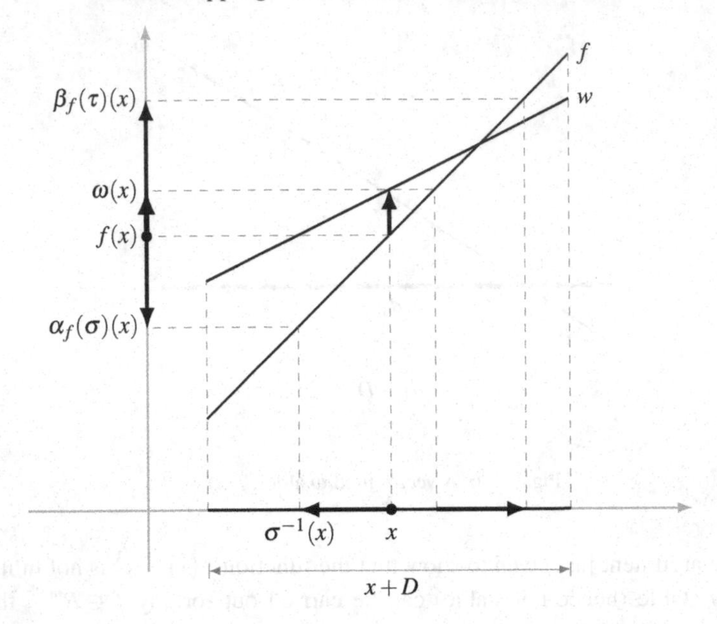

Figure 7.7 A vector field ω along $f(x) = x^2$

structure. This is the geometric essence of the Weierstrass-Malgrange-Mather preparation theorem [83], stated therein in algebraic form. The second step is to locally integrate certain vector fields that arise from a judicious application of the preparation theorem. Modulo these two 'jumps', the proof is rather simple and can be carried out entirely in the context of SDT. The preparation theorem holds in our choice model, as will be shown in the last part of this book. However, as we shall see in the next section, a different proof of Mather's theorem can be achieved within SDT without invoking the preparation theorem, for which reason we have not assumed it as part of SDT. We state it for use in this section as an *ad hoc* postulate.

Postulate 7.10 (Postulate PT) *Let $f \in R^{m\Delta(n)}$ and let $V \in W(R^{m\Delta(n)})$ where $f \in V$, that is, V is a neighbourhood of f for the weak topological structure. Let*

$$V \xrightarrow{\Phi} V^{[0,1]}$$

be any morphism such that $\Phi(f)(t) = f$ for all $t \in [0,1]$. If $d\gamma_{\Phi(f)}$ is surjective at $(\pi_{R^n} : R^n \times [0,1] \longrightarrow R^n, \pi_{R^m} : R^m \times [0,1] \longrightarrow R^m)$, it follows that $d\gamma_{\Phi|_{V^*}}$ is surjective at $(\pi_{R^n} : V^* \times R^n \times [0,1] \longrightarrow R^n, \pi_{R^m} : V^* \times R^m \times [0,1] \longrightarrow R^m)$, for some weak neighbourhood V^* such that $f \in V^* \subset V$.

Theorem 7.11 *Let $(\mathscr{E},R)$ be a model of SDT satisfying Postulate PT. Let $f \in R^{m\Delta(n)}$ with $f(0) = 0$. If f is infinitesimally stable, then f is stable.*

Proof Assume $f \in R^{m\Delta(n)}$ with $f(0) = 0$ is infinitesimally stable. The proof that it is stable will consist of five steps.
Step 1. Consider $\Phi\colon R^{m\Delta(n)} \times [0,1] \longrightarrow R^{m\Delta(n)}$ defined by

$$\Phi(g,t) = t \cdot g + (1-t) \cdot f$$

and notice that $\Phi(f,t) = f$ for any $t \in [0,1]$ and $\Phi(g,0) = f$ for any $g \in R^{\Delta(n)}$.

Let $V = V(\Delta(m), 1, f, \varepsilon)$ be given, for some $0 < \varepsilon < 1$, $\varepsilon \in R$. Thus, V is a basic open for the weak infinitesimal structure and $f \in V$. Since the weak topological structure is subintrinsic, V is an intrinsic open neighbourhood of f in $R^{m\Delta(n)}$.

We claim that the restriction of Φ to $V \times [0,1]$ has values in V. Indeed, if $g \in V$, by definition of V, we have

$$\forall x \in \Delta(n)\, [g(x) - f(x) \in (-\varepsilon, \varepsilon)]$$

and so, for any $t \in [0,1]$, by definition of V,

$$\forall x \in \Delta(n)\, [(t \cdot g(x) + (1-t) \cdot f(x) - f(x)) \in (-t\varepsilon, t\varepsilon)],$$

as the expression in parenthesis is just $t \cdot (g(x) - f(x))$. Now, $(-t\varepsilon, t\varepsilon) \subset (-\varepsilon, \varepsilon)$, as $t \in [0,1]$, and the claim is proved.

Thus, we may assume that we are dealing with

$$\Phi\colon V \longrightarrow V^{[0,1]}$$

with $f \in V$, $\Phi(f)(t) = f$ for all $t \in [0,1]$, and $\Phi(g)(0) = f$ for all g in V.

Furthermore, we also claim that, for *any* intrinsic neighbourhood $f \in V^* \subset V$ and $0 < \varepsilon' \leq 1$, $\Phi(V^* \times [0, \varepsilon'])$ is an intrinsic neighbourhood of f. In order to prove it, let $0 < t \leq \varepsilon'$ be arbitrary, and view $f = t \cdot f + (1-t) \cdot f \in V$. Given that $0 < t$, t is invertible and the following is meaningful and holds since, by assumption, V^* is an intrinsic neighbourhood of f contained in V:

$$\forall h \in V \big[\neg(t^{-1} \cdot (h - (1-t) \cdot f) = f) \vee t^{-1} \cdot (h - (1-t) \cdot f) \in V^*\big].$$

Now, this is the case if and only if

$$\forall h \in V \big[\neg(h = t \cdot f + (1-t) \cdot f) \vee h \in \Phi(V^* \times [0, \varepsilon'])\big],$$

that is, if and only if

$$\forall h \in V \big[\neg(h = \Phi(f)) \vee h \in \Phi(V^* \times [0, \varepsilon'])\big]$$

which proves the second claim.

Step 2. Since $\Phi(f) = f$, and f is infinitesimally stable, we can apply the preparation theorem (Postulate 7.10) to Φ. This gives some intrinsic neighbourhood V^* of f with $\alpha_{\Phi|_{V^*}} + \beta_{\Phi|_{V^*}}$ surjective. For the vector field

$$\omega = \frac{d\Phi|_{V^*}}{dt}$$

along $\Phi|_{V^*}\colon V^* \times [0,1] \times \Delta(n) \longrightarrow R^m$, this says that there exist two vector fields,
$\sigma \in \mathrm{Vect}\big(\pi_{R^n}\colon V^*\times[0,1]\times\Delta(n) \to R^n\big), \tau \in \mathrm{Vect}\big(\pi_{R^m}\colon V^*\times[0,1]\times\Delta(m) \to R^m\big)$
such that

$$\frac{d\Phi|_{V^*}}{dt} = \alpha_{\Phi|_{V^*}}(\sigma) + \beta_{\Phi|_{V^*}}(\tau)\,.$$

Step 3. In this step we shall apply Proposition 6.10 (which is a consequence of Postulate S) to (the principal parts of) the vector fields

$$g_\sigma \in R^{n V^*\times\Delta(n)\times[0,1]} \text{ and } g_\tau \in R^{m V^*\times\Delta(m)\times[0,1]}.$$

This gives uniquely the existence of

$$f_\sigma \in \Delta(n)^{V^*\times\Delta(n)\times\Delta} \text{ and } f_\tau \in \Delta(m)^{V^*\times\Delta(m)\times\Delta}$$

satisfying, for each $f^* \in V^*$, $x \in \Delta(n)$, $y \in \Delta(m)$ and $t \in \Delta$, the following two sets of conditions:

$$\begin{cases} f_\sigma(f^*,x,0) = x & x \in \Delta(n) \\ \dfrac{\partial f_\sigma}{\partial t}(f^*,x,t) = g_\sigma(f^*, f_\sigma(f^*,x,t),t) & t \in \Delta,\ x \in \Delta(n) \end{cases}$$

and

$$\begin{cases} f_\tau(f^*,y,0) = y & y \in \Delta(m) \\ -\dfrac{\partial f_\tau}{\partial t}(f^*,y,t) = g_\tau(f^*, f_\tau(f^*,y,t),t) & t \in \Delta,\ y \in \Delta(m) \end{cases}$$

Notice that, for $f = f^*$, since $\Phi(f,t)(x) = f(x)$, for each $x \in \Delta(n)$ and $t \in \Delta$, we may take σ, τ so that the corresponding g_σ and g_τ are such that

$$g_\sigma(f,x,0) = 0 \quad \forall t \in [0,1]\ \forall x \in \Delta(n)$$

and

$$g_\tau(f,y,0) = 0 \quad \forall t \in [0,1]\ \forall y \in \Delta(m).$$

Therefore the unique solutions f_σ and f_τ satisfy

$$\frac{\partial f_\sigma}{\partial t}(f,x,t) = 0 \qquad f_\sigma(f,x,0) = x$$

and

$$\frac{\partial f_\tau}{\partial t}(f,y,t) = 0 \qquad f_\tau(f,y,0) = y$$

which means that f_σ and f_τ do not depend on t, and then we must have

$$f_\sigma(f,x,t) = x \qquad \forall x \in \Delta(n)\, \forall t \in \Delta$$

and

$$f_\tau(f,y,t) = y \qquad \forall y \in \Delta(m)\, \forall t \in \Delta\,.$$

Then, the morphisms $f_\sigma(f,t)\colon \Delta(n) \to \Delta(n)$ and $f_\tau(f,t)\colon \Delta(m) \to \Delta(m)$ are the identity and, for every $f^* \in V^*$ and $t \in \Delta$, $f_\sigma(f^*,t)\colon \Delta(n) \longrightarrow \neg\neg\{x\}$ and $f_\tau(f^*,t)\colon \Delta(m) \longrightarrow \neg\neg\{y\}$ are both isomorphisms, where $x = f_\sigma(f,t)(0)$ and $y = f_\tau(f,t)(0)$.

Step 4. We now claim that, from $\dfrac{d\Phi_{|V^*}}{dt}(f^*,t) = \alpha_{\Phi_{|V^*}}(f^*,t) + \beta_{\Phi_{|V^*}}(f^*,t)$ it follows that $f \sim \Phi(f^*,t)$ as germs at x, that is, the diagram

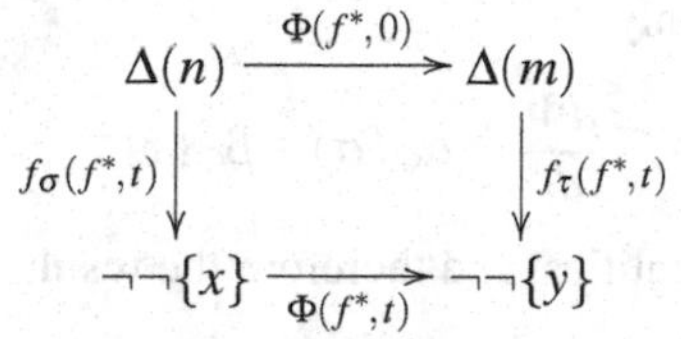

$$\begin{array}{ccc}
\Delta(n) & \xrightarrow{\Phi(f^*,0)} & \Delta(m) \\
{\scriptstyle f_\sigma(f^*,t)}\big\downarrow & & \big\downarrow{\scriptstyle f_\tau(f^*,t)} \\
\neg\neg\{x\} & \xrightarrow[\Phi(f^*,t)]{} & \neg\neg\{y\}
\end{array}$$

commutes, which would say that the deformation is trivial.

Reformulating the condition as $\Phi_0 = \Upsilon_t^{-1} \circ \Phi_t \circ \Psi_t$ where

$$\Psi = \langle \pi_1, \pi_2, f_\sigma \rangle\colon V^* \times [0,1] \times \Delta(n) \longrightarrow V^* \times [0,1] \times \Delta(n)\,,$$

and

$$\Upsilon = \langle \pi_1, \pi_2, f_\tau \rangle\colon V^* \times [0,1] \times \Delta(m) \longrightarrow V^* \times [0,1] \times \Delta(m)\,,$$

it now reads

$$\frac{d\Phi}{dt} = \alpha_\Phi\Big(\frac{d\Psi}{dt} \circ \Psi^{-1}\Big) + \beta_\Phi\Big(-\frac{d\Upsilon}{dt} \circ \Upsilon^{-1}\Big)$$

because $\sigma = \dfrac{d\Psi}{dt} \circ \Psi^{-1}$ and $\tau = -\dfrac{d\Upsilon}{dt} \circ \Upsilon^{-1}$.

We now have

$$\frac{d}{dt}(\Upsilon^{-1} \circ \Phi \circ \Psi) = \frac{d\Upsilon^{-1}}{dt} \circ \Phi \circ \Psi + (\Upsilon^{-1})^D \circ \frac{d\Phi}{dt} \circ \Psi + (\Upsilon^{-1})^D \circ \Phi^D \circ \frac{d\Psi}{dt}$$

and, from $\Upsilon^{-1} \circ \Upsilon = \mathrm{id}$ for any $t \in [0,1]$, it follows that $\frac{d}{dt}(\Upsilon^{-1} \circ \Upsilon) = 0$. Therefore $(\frac{d\Upsilon^{-1}}{dt} \circ \Upsilon) + (\Upsilon^{-1})^D \circ \frac{d\Upsilon}{dt} = 0$, and so $\frac{d\Upsilon^{-1}}{dt} = -(\Upsilon^{-1})^D \circ \frac{d\Upsilon}{dt} \circ \Upsilon^{-1}$, which gives

$$\begin{aligned}
\frac{d}{dt}(\Upsilon^{-1} \circ \Phi \circ \Psi) &= -(\Upsilon^{-1})^D \circ \frac{d\Upsilon}{dt} \circ \Upsilon^{-1} \circ \Phi \circ \Psi + (\Upsilon^{-1})^D \circ \frac{d\Phi}{dt} \circ \Psi \\
&\quad + (\Upsilon^{-1})^D \circ \Phi^D \circ \frac{d\Psi}{dt} \\
&= (\Upsilon^{-1})^D \circ \left[-\frac{d\Upsilon}{dt} \circ \Upsilon^{-1} \circ \Phi + \frac{d\Phi}{dt} + \Phi^D \circ \frac{d\Psi}{dt} \circ \Psi^{-1} \right] \circ \Psi \\
&= (\Upsilon^{-1})^D \circ \left[-\beta_\Phi(\tau) + \frac{d\Phi}{dt} + -\alpha_\Phi(\sigma) \right] \circ \Psi\,.
\end{aligned}$$

Since $(\Upsilon^{-1})^D$ and Ψ are isomorphisms, the last member of the equality vanishes if and only if $\dfrac{d\Phi}{dt} = \alpha_\Phi(\sigma) + \beta_\Phi(\tau)$. In other words, $\dfrac{d}{dt}(\Upsilon_t{}^{-1} \circ \Phi_t \circ \Psi_t) = 0$ if and only if $\dfrac{d\Phi}{dt} = \alpha_\Phi(\sigma) + \beta_\Phi(\tau)$. In particular,

$$\Upsilon_t{}^{-1} \circ \Phi_t \circ \Psi_t = \Upsilon_0{}^{-1} \circ \Phi_0 \circ \Psi_0$$

for all $t \in [0,1]$ if and only if

$$\frac{d\Phi}{dt} = \alpha_\Phi(\sigma) + \beta_\Phi(\tau)\,.$$

Now, Ψ_0 and Υ_0 are identities, and therefore the result is that

$$\Upsilon_t{}^{-1} \circ \Phi_t \circ \Psi_t = \Phi_0$$

if and only if

$$\frac{d\Phi}{dt} = \alpha_\Phi(\sigma) + \beta_\Phi(\tau).$$

□

Remark 7.12 (i) At first glance, the claimed application of the theorem on time-dependent systems (Proposition 6.10) in Step 3 of the proof of Theorem 7.11 may seem inappropriate. However, on the one hand, the V^* that occurs in the domain of the (principal part of the) vector field g_σ is just a parameter space and, on the other hand, by Axiom G, one can always extend $\Delta(n)$ to an open $M \subset R^n$ containing 0, and similarly extend $\Delta(m)$ to an open $N \subset R^m$ containing 0, so that what one applies Proposition 6.10 to is to each of the values of

$$g_\sigma : V^* \longrightarrow R^{nM \times [0,1]}$$

and

$$g_\tau \colon V^* \longrightarrow R^{N\times[0,1]}.$$

From Proposition 6.10 one then gets unique solutions

$$f_\sigma \colon V^* \longrightarrow M^{M\times\Delta}$$

and

$$f_\tau \colon V^* \longrightarrow N^{N\times\Delta},$$

which one can then restrict and get

$$f_\sigma \colon V^* \longrightarrow \Delta(n)^{\Delta(n)\times\Delta}$$

and

$$f_\tau \colon V^* \longrightarrow \Delta(m)^{\Delta(m)\times\Delta}.$$

(ii) In the same proof of Theorem 7.11, there appear some closed intervals of the form $[0,\varepsilon]$. As shown in [24] (Proposition II.3.5) Δ-flows can be extended uniquely to local flows on account of Axiom G. For a time-dependent system, and a Δ-flow $f \in R^{n\Delta(n)\times\Delta}$, it means that there is an extension $\bar{f} \in R^{nH}$ where $\Delta(n)\times\{0\} \subset H \in P(\Delta(n)\times R)$, $\bar{f}$ a flow. It follows that $H = \Delta(n)\times(-\varepsilon,\varepsilon)$ for some $0<\varepsilon'\leq 1$ and, for the flow equation to be satisfied, this will be so on $\Delta(n)\times((-\varepsilon,\varepsilon)\cap[0,1])$ hence, on $\Delta(n)\times[0,\varepsilon']$ for some $0<\varepsilon'\leq 1$.

(iii) Finally, if $f \in M^{M\times U}$ is a flow for $0 \in U \subset R$, the flow equation guarantees that $\bar{f} \in (M^M)^U$ is actually an element $\bar{f} \in (\mathrm{Iso}(M^M))^U$. Indeed, for any $t \in U$, $\bar{f}(t)^{-1}$ exists and is given by $\bar{f}(-t)$, provided that $-t \in U$, as follows from the identity

$$\bar{f}(x,0) = f(x,t+(-t)) = f(f(x,t),t).$$

This explains the last part of Step 3 in the proof of Theorem 7.11.

The notion of infinitesimal stability is, in our context, a logico-geometric notion as stated in Definition 7.5, itself based on the notion of infinitesimal surjectivity from Definition 7.1. The notion of infinitesimal surjectivity intervenes crucially in the proof of Theorem 7.6, in which it is shown that stability implies infinitesimal stability. Classically it is algebraic or analytic, as it refers to submersions and involves derivatives. However, as shown in Theorem 7.11, infinitesimal stability implies stability, which is a local notion, so that the two versions are equivalent also in our setting. Our advantage in SDT is to be able to use them both as stated (which differs from the classical counterparts) and as deemed appropriate in proofs or examples.

In the proof already given of the hard part of Mather's theorem ('infinitesimal stability implies stability') in SDT, an *ad hoc* postulate corresponding to the Malgrange preparation theorem was used. We now show that an alternative proof can be given without that extra assumption. This comes at a price, which is to introduce yet other forms of stability, namely V-infinitesimal stability, almost V-infinitesimal stability, and transversal stability. The V-equivalence of germs allows us to use elementary algebraic techniques, similar to those of analytic geometry. Moreover, the notion of infinitesimal stability can easily be translated into a transversality condition that is the key to completing the proof. We begin with some remarks that are intended to serve as motivation for the new notions adopted in this chapter.

Let $(\mathscr{E},R)$ be a model of SDT. Let $f \in R^{m\Delta(n)}$. Denote by $\mathfrak{M}_{m\times m}(R)$ the object of $m \times m$ matrices with entries in R, and by $GL(m)$ its subobject consisting of the invertible matrices. Consider the group

$$G' = \mathrm{Inv}\big(\Delta(n)^{\Delta(n)}\big) \times \mathrm{GL}(m)^{\Delta(n)} \times R^m$$

and define

$$\Gamma_f \colon G' \longrightarrow R^{m\Delta(n)}$$

by means of

$$\Gamma_f(\varphi,A,b)(x) = A(x)\cdot f(\varphi^{-1}(x)) + b\,.$$

If a germ $g \in R^{m\Delta(n)}$ is in the G'-orbit of f we say that f and g are *V-equivalent.*

Definition 7.13 A germ $f \in R^{m\Delta(n)}$ is said to be *V-infinitesimally stable* if Γ_f is a submersion at $(\mathrm{id}_n, \bar{\mathrm{I}}_m, 0)$, where $\bar{\mathrm{I}}_m(x) = \mathrm{I}_m \in \mathfrak{M}_{m\times m}(R)$ for any $x \in \Delta(n)$.

Theorem 7.14 *A germ $f \in R^{m\Delta(n)}$ is infinitesimally stable if and only if it is V-infinitesimally stable.*

Proof 1. The necessity part will be a consequence of finding a morphism

$$\Theta \colon T_{(\mathrm{id}_n,\mathrm{id}_m)}G \longrightarrow T_{(\mathrm{id}_n,\bar{\mathrm{I}}_m,0)}G'$$

such that the following diagram commutes.

$$\begin{array}{ccc} T_{(\mathrm{id}_n,\mathrm{id}_m)}G & \xrightarrow{\Theta} & T_{(\mathrm{id}_n,\bar{\mathrm{I}}_m,0)}G' \\ & {\scriptstyle (d\gamma_f)_{(\mathrm{id}_n,\mathrm{id}_m)}} \searrow \quad \swarrow {\scriptstyle (d\Gamma_f)_{(\mathrm{id}_n,\bar{\mathrm{I}}_m,0)}} & \\ & T_f(R^{m\Delta(n)}) & \end{array}$$

To this end, given

$$(\sigma,\tau)\in T_{(\mathrm{id}_n,\mathrm{id}_m)}G\subset \Delta(n)^{\Delta(n)\times D}\times R^{mR^m\times D}$$

consider the principal part function k of the vector field τ, that is, $k\in R^{mR^m}$ such that for all $d\in D$ and $y\in R^m$, $\tau(y,d)=y+d\cdot k(y)$.

It follows easily from Postulate F (Postulate 2.5) that for this k there exists a unique

$$C\colon R^m\times R^m \longrightarrow \mathfrak{M}_{m\times m}(R)$$

such that for every $y\in R^m$,

$$k(y)-k(0)=C(y,0)\cdot y\,.$$

With this C we can now define a vector field μ along $\bar{\mathrm{I}}_m$ by means of $\mu(x,d)=\bar{\mathrm{I}}_m(x)+d\cdot C(f(x),0)=\mathrm{I}_n+d\cdot C(f(x),0)$ and notice that this $\mu(x,d)$ belongs to $GL(m)$ as every infinitesimal deformation of the identity matrix I_n is invertible. Therefore, $\mu\in T_{\bar{\mathrm{I}}_m}(GL(m)^{\Delta(n)})$. Let η be given by $\eta(d)=d\cdot k(0)$. We have $\eta\in T_0R^m$. Letting $\Theta(\sigma,\tau)=(\sigma,\mu,\eta)$ gives what was wanted.

2. The sufficiency part is proven in a similar way to produce a morphism

$$\Lambda\colon T_{(\mathrm{id}_n,\bar{\mathrm{I}}_m,0)}G' \longrightarrow T_{(\mathrm{id}_n,\mathrm{id}_m)}G$$

such that $(d\gamma_f)_{(\mathrm{id}_n,\mathrm{id}_m)}\circ\Lambda=(d\Gamma_f)_{(\mathrm{id}_n,\bar{\mathrm{I}}_m,0)}$.

Given $(\sigma,\mu,\eta)\in T_{(\mathrm{id}_n,\bar{\mathrm{I}}_m,0)}G'$, let $A\in(\mathfrak{M}_{m\times m}(R))^{\Delta(n)}$ and $b\in R^m$ denote the principal parts of μ and η, respectively. Define

$$\tau(f(x),d)=f(x)+d\cdot(A(x)\cdot f(x))+b$$

and then the map Λ, given by $\Lambda(\sigma,\mu,\eta)=(\sigma,\tau)$, can be checked to be as required. □

Among the infinitesimal deformations of f induced by

$$\Gamma_f\colon G' \longrightarrow R^{m\Delta(n)}\,,$$

that is, among those in the image of $d\Gamma_{f\,(\mathrm{id}_n,\bar{\mathrm{I}}_m,0)}$, some correspond to

$$T_{(\mathrm{id}_n,\bar{\mathrm{I}}_m,0)}\Big(\mathrm{Inv}\big(\Delta(n)^{\Delta(n)}\big)\times \mathrm{GL}(m)^{\Delta(n)}\times 0\Big)\,,$$

the trivial deformations.

Proposition 7.15 *Let $f\in R^{m\Delta(n)}$ be V-infinitesimally stable and let $M\subset$ $\mathrm{Vect}f$ be the subobject of trivial deformations of f. Then $\mathrm{indep}_R(\mathrm{Vect}f/_M)\leq$ m.*

Proof Implicit in the diagram below is the fact that $T_0R^m \simeq R^m$. If f is V-infinitesimally stable, the middle horizontal arrow in the diagram below is an epimorphism.

$$\begin{array}{ccc}
T_{\mathrm{id}_m}(\Delta(n)^{\Delta(n)}) \times T_{\bar{\mathrm{I}}_m}(\mathrm{GL}(m)^{\Delta(n)}) & \xrightarrow{(d\Gamma_f)_{(\mathrm{id}_n,\bar{\mathrm{I}}_m)}} & M \\
\downarrow & & \downarrow \\
T_{\mathrm{id}_m}(\Delta(n)^{\Delta(n)}) \times T_{\bar{\mathrm{I}}_m}(\mathrm{GL}(m)^{\Delta(n)}) \times R^m & \xrightarrow{(d\Gamma_f)_{(\mathrm{id}_n,\bar{\mathrm{I}}_m,0)}} & \mathrm{Vect}f \\
\downarrow & & \downarrow \\
R^m & \xrightarrow{\zeta} & \mathrm{Vect}f/M
\end{array}$$

It follows from this then that the bottom horizontal arrow ζ is an epimorphism. Since $\mathrm{indep}_R(R^m) \leq m$, by Lemma 2.41, we get $\mathrm{indep}_R(\mathrm{Vect}f/M) \leq m$, which is what we wanted to prove. □

We now show that, for a V-infinitesimally stable germ in $R^{m\Delta(n)}$, there are no infinitesimal deformations of f of degree bigger than m that are independent of the trivial ones.

In what follows we consider the ideal of $R^{\Delta(n)}$ given by

$$\mathfrak{m} = [[\varphi \in R^{\Delta(n)} \mid \varphi(0) = 0]] \,.$$

Lemma 7.16 *Let $f \in R^{m\Delta(n)}$ be a V-infinitesimally stable (or, equivalently, infinitesimally stable) germ. Denote by $M \subset \mathrm{Vect}f$ the subobject of infinitesimal deformations of f. Then $\mathfrak{m}^m\mathrm{Vect}f \subset \neg\neg M$.*

Proof The statement of the lemma follows from Theorem 2.40 and Proposition 7.15. In Theorem 2.40 let $A = R$ and $V = \mathrm{Vect}f$, so that $M \subset V$ as an R-submodule. We need to remark that $\mathrm{Vect}f$ is finitely generated as an $R^{\Delta(n)}$-module—indeed, it is generated by the m infinitesimal deformations $\hat{e}_i(x,d) = f(x) + d \cdot e_i$ of f, with $\{e_1, \ldots, e_m\}$ the canonical basis of R^m. □

As observed earlier, the condition, in terms of the principal parts functions, for a germ $f \in R^{m\Delta(n)}$ to be V-infinitesimally stable, reduces to the solvability of the equation

$$g(x) = -\frac{df}{dx}(x)h(x) + A(x) \cdot f(x) + b$$

for $h \in R^{n\Delta(n)}$, $A \in (\mathfrak{M}_{m\times m}(R))^{\Delta(n)}$, and $b \in R^m$, for every germ $g \in R^{m\Delta(n)}$.

If we require this equation to be solvable *only* at the level of jets, then we obtain a new kind of stability. To this end, consider the restriction map

$$j_0^m : R^{m\Delta(n)} \longrightarrow R^{mD_m(n)} .$$

Definition 7.17 A germ $f \in R^{m\Delta(n)}$ is said to be *almost V-infinitesimally stable* if the composite

$$G' \xrightarrow{\Gamma_f} R^{m\Delta(n)} \xrightarrow{j_0^m} R^{mD_m(n)}$$

is a submersion at $(\mathrm{id}_n, \bar{\mathrm{I}}_m, 0)$.

The condition of almost *V*-infinitesimal stability can be rephrased in terms of the existence of solutions to the equation corresponding to *V*-infinitesimal stability but for *m*-jets, that is, the solvability of the same equation but for $D_m(n)$.

Explicitly, a germ $f \in R^{m\Delta(n)}$ is almost *V*-infinitesimally stable if and only if for each $g \in R^{m\Delta(n)}$ there exist $h \in R^{n\Delta(n)}$, $A \in (\mathfrak{M}_{m\times m}(R))^{\Delta(n)}$ and $b \in R^m$ such that for each $t \in D_m(n)$,

$$g(t) = f(t) + d \cdot \left(-\frac{df}{dx}(t)h(t) + A(t) \cdot f(t) + b \right) .$$

It is obvious that if a germ is *V*-infinitesimally stable then it is also almost *V*-infinitesimally stable. In a way analogous to the proof of Lemma 7.16 for *V*-infinitesimally stable germs, and with the same notation as therein, we now get the following.

Lemma 7.18 *Let $f \in R^{m\Delta(n)}$ be almost V-infinitesimally stable. Then,*

$$\mathfrak{m}^m \mathrm{Vect} f \subset \neg\neg M .$$

In what follows we shall establish that the condition on the *m*-jets is sufficient for *V*-infinitesimal stability. Consider the morphism

$$(d\, j_0^k)_f \colon \mathrm{Vect} f \longrightarrow \mathrm{Vect} f|_{D_k(n)}$$

defined by $(d\, j_0^k)_f(\omega) = \omega|_{D_k(n)}$ [1]

$$\mathfrak{m}^{k+1}\mathrm{Vect} f = [[\sum_{i=1}^{r} \varphi_i\, \omega_i \mid \varphi_i \in \mathfrak{m}^{k+1} \text{ and } \omega_i \in \mathrm{Vect} f, \text{ for } i = 1, \ldots, r]] .$$

Indeed, if $\omega \in \mathrm{Ker}(d\, j_0^k)_f$ and $g \in R^{m\Delta(n)}$ its principal part, say $\omega(d) = f + d \cdot g$ for each $d \in D$, it must be the case that $\omega(d)|_{D_k(n)} = f|_{D_k(n)}$, and this

[1] Note that $(d\, j_0^k)_f(\mathrm{Vect} f) = (\mathrm{Vect} f)|_{D_k(n)} = \mathrm{Vect} f|_{D_k(n)} = \mathrm{Vect}\, j_0^k f$.

is equivalent to $g|_{D_k(n)} = 0$. Since $g = (g_1, \ldots, g_m)$, we must have that $\omega \in \mathrm{Ker}(d\, j_0^k)_f$ if and only if $g_i \in \mathfrak{m}^{k+1}$ for each $i = 1, \ldots, m$. Therefore, if $\omega \in \mathrm{Ker}(d\, j_0^k)_f$ then $\omega = \sum_{i=1}^m g_i \hat{e}_i$ with $g_i \in \mathfrak{m}^{k+1}$ and so $\omega \in \mathfrak{m}^{k+1}\mathrm{Vect}f$.

Let $\sum_{i=1}^r \varphi_i \omega_i \in \mathfrak{m}^{k+1}\mathrm{Vect}f \subset \mathrm{Vect}f$, then its principal part is $\sum_{i=1}^r \varphi_i\, \bar{\omega}_i$, where $\bar{\omega}_i$ denotes the principal part of each ω_i. Since each $\varphi_i \in \mathfrak{m}^{k+1}$, all summands are in $\mathfrak{m}^{k+1} R^{m\Delta(n)}$ and so $\sum_{i=1}^r \varphi_i\, \omega_i \in \mathrm{Ker}(d\, j_0^k)_f$.

It follows from this that

$$\mathrm{Vect}f|_{D_k(n)} \simeq \mathrm{Vect}f/\mathfrak{m}^{k+1}\mathrm{Vect}f\,.$$

Theorem 7.19 *If $f \in R^{m\Delta(n)}$ is almost V-infinitesimally stable, then f is V-infinitesimally stable.*

Proof If f is almost V-infinitesimally stable then

$$\mathfrak{m}^m\mathrm{Vect}f \subset \neg\neg M\,.$$

In addition,

$$\begin{aligned}\mathrm{Vect}f/\mathfrak{m}^{m+1}\mathrm{Vect}f &= \mathrm{Im}\,(d\, j_0^m \circ \Gamma_f)_{(\mathrm{id}_n, \bar{1}_m, 0)}\\ &= \mathrm{Im}\,(d\,\Gamma_f)_{(\mathrm{id}_n, \bar{1}_m, 0)}/\mathfrak{m}^{m+1}\mathrm{Vect}f.\end{aligned}$$

Therefore,

$$\begin{aligned}\mathrm{Vect}f &= \mathrm{Im}\,(d\Gamma_f)_{(\mathrm{id}_n, \bar{1}_m, 0)} + \mathfrak{m}^{m+1}\mathrm{Vect}f\\ &\subset \neg\neg\mathrm{Im}\,(d\Gamma_f)_{(\mathrm{id}_n, \bar{1}_m, 0)} + M\\ &\subset \neg\neg\mathrm{Im}\,(d\Gamma_f)_{(\mathrm{id}_n, \bar{1}_m, 0)}.\end{aligned}$$

Thus, $\mathrm{Vect}f = \mathrm{Im}\,(d\,\Gamma_f)_{(\mathrm{id}_n, \bar{1}_m, 0)}$ and f is V-infinitesimally stable. □

As a consequence of this result we obtain the following.

Corollary 7.20 *Let $f, g \in R^{m\Delta(n)}$ be such that $f|_{D_{m+1}(n)} = g|_{D_{m+1}(n)}$. If f is infinitesimally stable then so is g.*

Definition 7.21 A germ $f \in R^{m\Delta(n)}$ is said to be *k-determined* if for any $g \in R^{m\Delta(n)}$ such that $f|_{D_k(n)} = g|_{D_k(n)}$, one has $g \in \mathrm{Im}\,(\gamma_f)$.

We now come to a result that will make clear that the property of infinitesimal stability for a germ $f \in R^{m\Delta(n)}$ is a property of its $(m+1)$-jet and this gives a normal form for any infinitesimally stable germ. The method used to prove it is the homotopical method of R. Thom [106, 108].

Theorem 7.22 *Let $f \in R^{m\Delta(n)}$ be infinitesimally stable. Then, f is $(m+1)$-determined.*

Proof Let $\varphi \in \mathfrak{m}^{m+2}R^{m\Delta(n)}$. We will show that $f+\varphi$ is equivalent to f. To that end, we join f to $f+\varphi$ by a path $f_t = f + t\cdot\varphi$, $t \in [0,1]$, and we will see that each f_t is equivalent to f. It will be enough to find diffeomorphisms

$$H\colon \Delta(n)\times[0,1] \longrightarrow \Delta(n)$$

$$K\colon \Delta(m)\times[0,1] \longrightarrow \Delta(m)$$

such that for each $t \in [0,1]$, $K_t \circ f \circ H_t{}^{-1} = f_t$ where

$$\begin{array}{ccc} \Delta(n)\times[0,1] & \xrightarrow{f\times \mathrm{id}} & \Delta(m)\times[0,1] \\ \Big\downarrow{\scriptstyle\langle H,\pi_2\rangle} & & \Big\downarrow{\scriptstyle\langle K,\pi_2\rangle} \\ \Delta(n)\times[0,1] & \xrightarrow[\langle F,\pi_2\rangle]{} & \Delta(m)\times[0,1] \end{array}$$

where

$$\langle F,\pi_2\rangle(x,t) = (f_t(x),t)$$
$$\langle H,\pi_2\rangle(x,t) = (H(x,t),t)$$
$$\langle K,\pi_2\rangle(x,t) = (K(x,t),t)\,.$$

We will obtain H and K as the integral curves of vector fields h_t and k_t depending on time from the equations

$$H_0(x) = x \qquad \frac{dH_t}{dt}(x) = h_t(H_t(x))$$
$$K_0(y) = y \qquad \frac{dK_t}{dt}(y) = k_t(K_t(y))\,,$$

which we derive to obtain

$$\varphi(H_t(x)) + \frac{df_t}{dx}(H_t(x))h_t(H_t(x)) = k_t(f_t(H_t(x)))\,.$$

Since this equation must hold for every $x \in \Delta(n)$ and $t \in [0,1]$, we can write

$$\varphi(x) = -\frac{df_t}{dx}(x)h_t(x) + k_t(f_t(x))\,.$$

We can think of φ as the principal part of a vector field along $f_t = f + t\cdot\varphi$. Now, f and f_t have the same $(m+1)$-jet and so, f_t too is infinitesimally stable and we can thus find h_t and k_t that solve the equation for each $t \in [0,1]$. To finish the proof it will be enough to shown that for each $t \in [0,1]$, $h_t \in \Delta(n)^{\Delta(n)}$

and $k_t \in \Delta(m)^{\Delta(m)}$, as this will give that $H_t \in \Delta(n)^{\Delta(n)}$ and $K_t \in \Delta(m)^{\Delta(m)}$, both invertible by the very condition of integral flow. Now, this follows from the solutions to the equation for $\varphi \in \mathfrak{m}^m R^{m\Delta(n)}$ given by the infinitesimal stability of the f_t, that is, from

$$\varphi(x) = -\frac{df_t}{dx}(x)h_t(x) + k_t(f_t(x)),$$

which forces $\neg\neg(k_t(0) = 0)$, as $\mathfrak{m}^m \mathrm{Vect} f \subset \neg\neg M$, and from this $\neg\neg(h_t(0) = 0)$. □

Remark 7.23 The solvability of the equation corresponding to infinitesimal stability at the level of jets for a germ $f \in R^{m\Delta(n)}$ (that is, the condition that f is almost infinitesimally stable) admits, in the framework of SDT, an interesting geometric interpretation. This is also the case for the condition of almost V-infinitesimal stability, which means that $j_0^m \circ \Gamma_f$ is a submersion at $(\mathrm{id}_n, \bar{1}_m, 0)$ or, equivalently, after Theorem 7.14, $j_0^m \circ \gamma_f$ is a submersion at $(\mathrm{id}_n, \mathrm{id}_m)$.

We now come to a key definition that will allow us to obtain the implication 'infinitesimal stability implies stability'. This is a new notion of 'transversal stability'.

Let $\mathcal{O}_k f$ denote the orbit of the k-jet of f, that is, the image of $j_0^k \circ \gamma_f$, and consider

$$J^k f \colon \Delta(n) \longrightarrow R^{mD_k(n)}$$

defined by $J^k f(x)(t) = f(x+t)$ for $x \in \Delta(n)$, $t \in D_k(n)$.

Definition 7.24 A germ $f \in R^{m\Delta(n)}$ is said to be *k-transversally stable* if $J^k f \pitchfork_0 \mathcal{O}_k f$. Equivalently, the condition says that

$$T_{f|_{D_k(n)}}(R^{mD_k(n)}) = \mathrm{Im}(dJ^k f)_0 + T_{f|_{D_k(n)}}(\mathcal{O}_k f)\,.$$

Theorem 7.25 *A germ $f \in R^{m\Delta(n)}$ is almost V-infinitesimally stable if and only if it is k-transversally stable for all $k \geq m$.*

Proof Assume that f is k-transversally stable, where $k \geq m$. To get the conclusion that f is almost V-infinitesimally stable, it is enough, by the proof of Theorem 7.14, to prove that the differential at $(\mathrm{id}_n, \mathrm{id}_m)$ of the composite $j_0^k \circ \gamma_f$ is a submersion. To this end, let us consider $\omega \in T_{f|_{D_k(n)}}(R^{mD_k(n)})$. By assumption there exist $v \in \mathrm{Im}\,(dJ^k f)_0$ and $\chi \in T_{f|_{D_k(n)}}(\mathcal{O}_k f)$ such that $\omega = v + \chi$.

Since $v \in \mathrm{Im}\,(dJ^k f)_0$, its principal part is of the form $v \cdot \frac{df|_{D_k(m)}}{dx}$. Now, given that $\chi \in T_{f|_{D_k(n)}} \mathcal{O}_k f$, there exist $\bar{\sigma} \in T_{\mathrm{id}_n}(\Delta(n)^{\Delta(n)})$ and $\tau \in T_{\mathrm{id}_n}(R^{mR^m})$ such that $d(\bar{\sigma},\tau)_{(\mathrm{id}_n,\mathrm{id}_m)} = \chi$. Since $\bar{\sigma} \in T_{\mathrm{id}_n}(\Delta(n)^{\Delta(n)})$, it follows that for each $d \in D$ and $x \in \Delta(n)$, $\bar{\sigma}(d)(x) = x + d \cdot h(x)$. Define $\sigma(d)(x) = x + d \cdot (h(x) - v)$. Clearly $\sigma \in T_{\mathrm{id}_n}(\Delta(n)^{\Delta(n)})$. It is easy to verify that

$$d(j_0^k \circ \gamma_f)_{(\mathrm{id}_n,\mathrm{id}_m)}(\sigma,\tau) = \omega\,.$$

The converse is easily verified, in fact, for any k. □

Lemma 7.26 *Let $f, f^* \in R^{m\Delta(n)}$, f infinitesimally stable and $f^*|_{D_{m+1}(n)} \in \mathscr{O}_{m+1}f$, then f^* is equivalent to f.*

Proof If $f^*|_{D_{m+1}(n)} \in \mathscr{O}_{m+1}f$, then from the definition of $\mathscr{O}_{m+1}f$ it follows that $f^*|_{D_{m+1}(n)} = \psi \circ f \circ \varphi^{-1}|_{D_m(n)}$ for some $(\varphi, \psi) \in G$. By Theorem 7.22, f^* is equivalent to $\psi \circ f \circ \varphi^{-1}$ which, in turn, is equivalent to f. □

We arrive now to the main theorem of this section.

Theorem 7.27 *Let $(\mathscr{E}, R)$ be a model of SDT. Let $f \in R^{m\Delta(n)}$ with $f(0) = 0$. If f is infinitesimally stable, then f is stable.*

Proof By Theorem 7.25, if f is infinitesimally stable, then it is $(m+1)$-transversally stable, that is $J^{m+1}f \pitchfork_0 \mathscr{O}_{m+1}f$. By Proposition 6.36, we have

$$\neg\neg\{f|_{D_{m+1}(n)}\} \cap \mathscr{O}_{m+1}f = g^{-1}\{0\}$$

where $g = \langle g_1, \ldots, g_s \rangle$ with $g_1, \ldots, g_s$ independent functions. Using that the object of submersions is a weak open, we can find a weak open V such that

$$f \in V \subset [[h \in R^{m\Delta(n)} \mid g \circ J^{m+1}h \in \mathrm{Subm}_0]]\,.$$

Every $f^* \in V$ meets $\mathscr{O}_{m+1}f$ transversally at 0 and therefore $f^*|_{D_{m+1}(n)} \in \mathscr{O}_{m+1}f$. Thus, f^* is equivalent to f, as required. □

8
Morse Theory in SDT

One of the main uses of Mather's theorem is the classification of singularities of smooth mappings between real manifolds in low dimensions [13, 47]. The main portion of this chapter is an application of Mather's theorem within SDT to Morse theory [44, 45, 46].

8.1 Generic Properties of Germs in SDT

Let $(\mathscr{E}, R)$ be a model of SDT. Recall Definition 7.2 —the definition of equivalence $g \sim f$ for germs f, g of mappings $R^n \longrightarrow R^m$ in $\mathscr{E}$ regarded internally as elements of the object $R^{m\Delta(n)}$ by virtue of Axiom G. Recall also Definition 7.4 of a germ being stable and of the group G of diffeomorphisms that is part of the definition.

Definition 8.1 We say that a subobject

$$\Phi \subset R^{p\Delta(n)}$$

corresponds to a *generic property* of germs if

(i) $\Phi \subset R^{p\Delta(n)}$ is dense for the intrinsic (or Penon) topological structure, that is, the following statement

$$\forall U \in P\big(R^{p\Delta(n)}\big)\ \exists f \in (U \cap \Phi)$$

holds in $\mathscr{E}$, and

(ii) Φ is closed under the action of the group G, that is

$$\forall f, g \in R^{p\Delta(n)}\ [(f \in \Phi \wedge g \sim f) \Rightarrow g \in \Phi].$$

Theorem 8.2 *For any generic property Φ of $R^{p\Delta(n)}$, any stable germ $f \in R^{p\Delta(n)}$ has property Φ.*

Proof Let $f \in R^{p\Delta(n)}$ be a stable germ. Then, $\operatorname{Im}\gamma_f \subset R^{p\Delta(n)}$ is weak open hence intrinsic open. Apply (1) in the definition of a generic property to $U = \operatorname{Im}\gamma_f$. Since Φ is dense, there exists $g \in \operatorname{Im}\gamma_f$ such that $g \in \Phi$. Now, $g \sim f$ by definition of γ_f, and $g \in \Phi$, so that by (2) in the definition of generic, $f \in \Phi$. □

An impossible goal of the classification of singularities is to find, for any given pair (n, p) of positive integers, a finite list $\mathscr{L}_{(n,p)}$ of generic properties of germs such that, for any $f \in R^{p\Delta(n)}$, f is stable if and only if it satisfies the generic properties in $\mathscr{L}_{n,p}$. A reason why this is so is that, in general, since stable germs in $R^{p\Delta(n)}$ are finitely determined, hence stability is determined by their $(p+1)$-jets, the property $\Phi_{\mathscr{O}}$ defined by

$$f \in \Phi_{\mathscr{O}} \Leftrightarrow J^{p+1}f \pitchfork_0 \mathscr{O}_{p+1}f\,,$$

where $\mathscr{O}_{p+1}$ is an orbit under the action of G, is generic, yet it does not translate into a reasonable condition, let alone be reducible to a finite list. For this reason one can only hope for the classification problem to be tractable just for certain pairs (n, m) of dimensions.

Classically, there is a list of good pairs of dimensions for which a complete classification can be given. We give an example—to wit, immersions with normal crossings in good pairs of dimensions. The discussion takes place in SDT.

Consider $f \in R^{pR^n}$, with $p \geq 2n$. Assume that f is an immersion.

Definition 8.3 Say that $f \in Y^X$ is *an immersion with normal crossings* if it is an immersion and for each $s > 1$,

$$f^{[s]} \pitchfork \operatorname{diag}(Y^s)\,,$$

where $f^{[s]} : X^{[s]} \longrightarrow Y^{[s]}$ is the restriction of $f^s : X^s \longrightarrow Y^s$ to

$$X^{[s]} = [[(x_1, \ldots, x_s) \in X^s \mid \bigwedge_{1 \leq i < j \leq s} \neg(x_i = x_j)]]$$

and where

$$\operatorname{diag}(Y^s) = [[(y, \ldots, y) \in Y^s \mid y \in Y]]\,.$$

Let Φ be the corresponding property of germs in R^{pR^n}. It was stated in Proposition 6.30 that immersions are dense for the weak topology, which is subintrinsic. The condition defining the class Φ is that of transversal stability, hence equivalent to stability. Therefore, the property Φ is generic, so that for $p \geq 2n$, the stable germs are precisely the immersions with normal crossings.

We remark that not all immersions are stable—for instance, in the case R^R

(which does not satisfy the condition $p \geq 2n$ as $p = n = 1$), the figure below shows that small deformations destroy equivalence since the number of self-intersections is an invariant of equivalence.

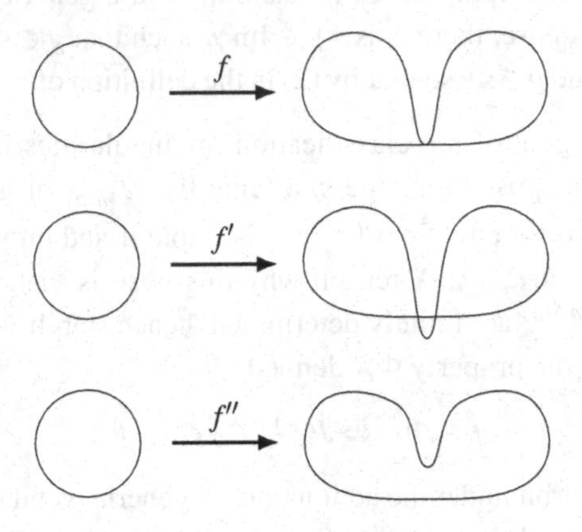

Figure 8.1 Non-stable immersion

In the next section, we shall give another example of a generic property, namely Morse germs.

8.2 Morse Germs in SDT

Let $(\mathscr{E}, R)$ be a model of SDT,

Definition 8.4 Let $x_0 \in R^n$, $X = \neg\neg\{x_0\}$, and $f \in R^X$. We say that $x \in X$ is a *singularity* of f if f is constant on $x + D(n)$.

If we consider the subobject of constant 1-jets at 0

$$S_1 = [[g \in R^{D(n)} \mid \forall \delta \in D(n)\, (g(\delta) = g(0))]]$$

and define $J^1 f : X \longrightarrow R^{D(n)}$ as $J^1 f(x)(\delta) = f(x + \delta)$, for $\delta \in D(n)$, the condition for x to be a singularity of f is that $J^1 f(x) \in S_1$.

Notice that S_1 is a submanifold of $R^{D(n)}$. Indeed, every value of a submersion is a regular value, and the Preimage Theorem (Corollary 6.23) gives that S_1 is a submanifold of $R^{D(n)}$ of codimension 1.

Definition 8.5 Let $x_0 \in R^n$, $X = \neg\neg\{x_0\}$, and $f \in R^X$. A singularity x of f is said to be *non-degenerate* if $J^1 f \pitchfork_x S_1$.

In the presence of the axiom of jets representability (Axiom 2.1), $J^1f(x) \in D^{D(n)}$ can be identified with $\left(f(x), \frac{\partial f}{\partial x_1}(x), \ldots, \frac{\partial h}{\partial x_n}(x)\right) \in R \times R^n$, and the following result gives an internal characterization of non-degenerate singularities.

Proposition 8.6 *Let $f \in R^X$. An element $x \in X$ is a non-degenerate singularity of f if and only if $\pi \circ J^1 f$ is a submersion at x, where $\pi\colon R \times R^n \longrightarrow R^n$ is the canonical projection.*

Proof The condition that $x \in X$ is a non-degenerate singularity of f translates into the equation

$$T_{J^1f(x)}R^{D(n)} = \mathrm{Im}\left[(dJ^1f)_x\right] + T_{J^1f(x)}S_1 \,.$$

Since the functor $(-)^D$ has a left adjoint, it preserves products, hence π^D is itself a projection. We have that

$$\forall \tau \in T_{J^1f(x)}(R \times R^n)\,(\tau \in T_{J^1f(x)}S_1 \Leftrightarrow \pi^D(\tau) = 0)\,.$$

In particular, for any non-degenerate singularity x of f, we have that

$$T_{J^1f(x)}R^{D(n)} = \mathrm{Im}\left[(dJ^1f)_x\right] + \mathrm{Ker}\left[(d\pi)_{J^1f(x)}\right].$$

It follows from the characterization of submersions that $\pi \circ J^1 f$ is a submersion at x if and only if $d(\pi \circ J^1 f)_x$ is locally surjective. Now, π^D, being a projection, is a submersion and therefore (actually equivalently) $d\pi_{J^1f(x)}$ is locally surjective, hence the desired result follows. □

Corollary 8.7 *Let $f \in R^X$, $X = \neg\neg\{x_0\} \subset R^n$. If $x \in X$ is a non-degenerate singularity of f, then the Hessian of f at x, that is, the matrix*

$$\left(\frac{\partial^2 f}{\partial x^2}(x)\right)$$

is non-singular.

Proof By Proposition 8.6, if x is a non-degenerate singularity of f, then the map $\pi \circ J^1 f$ is a submersion at x. The result now follows from Proposition 6.14 (ii), since the corresponding set of vectors is precisely the set of rows of the Hessian at x. □

Definition 8.8 A germ $f \in R^X$ for $X = \neg\neg\{x_0\} \subset R^n$ is said to be a *Morse germ* if the following statement holds in $\mathscr{E}$:

$$\forall x \in X\left[J^1f(x) \in S_1 \Rightarrow J^1 f \pitchfork_x S_1\right].$$

Remark 8.9 An important use of Morse functions (or germs) is towards an analysis of the behaviour of a manifold at a given point. For this reason, it is useful to know whether there are any Morse germs at all, and by how much does a given germ differ from one that has this property.

The following result answers the second question of Remark 8.9.

Proposition 8.10 *The subobject of Morse germs in R^X is dense in R^X for the weak topological structure.*

Proof Recall that $f \in R^X$ is a Morse germ if and only if $J^1 f \pitchfork S_1$. The result follows from Thom's transversality theorem (Theorem 6.39) together with the observations that $S_1 \subset R^{D(n)}$ is a submanifold cut out by independent functions. □

One of the basic results of Morse theory in characterizing the behaviour of manifolds at singular points says that the non-degenerate singularities are isolated.

Proposition 8.11 *A Morse germ has at most one singularity.*

Proof Let $f \in R^X$ be a Morse germ. If $x \in X$ is a singularity, it is non-degenerate and so $\pi_1 \circ J^1 f$ is a submersion at x and 0 is a regular value for this map. From the corresponding version for germs of Corollary 6.23 we have

$$[[x \in X \mid x \text{ is a singularity of} f]] = (\pi \circ J^1 f)^{-1}(0) \subset X$$

is a submanifold. Moreover, it has codimension n, hence dimension 0. □

Remark 8.12 The restriction to germs with codomain R in the theory of singularities can easily be lifted and the results extended to germs with codomain R^m for arbitrary m. In our set-up, the object S_1 of singularities of a given germ $f \in R^X$ has codimension 1. Its definition can be extended in the case of germs R^{mX} to S_r, for $r \leq m$. The proof that these objects are submanifold of the corresponding jet space is the fundamental result of Thom-Boardman stratification theory [48], and techniques such as that of identifying Morse germs as those which are transversal to the corresponding object of singularities are one of the tools employed towards the classification of singularities [47].

We now address the first question of Remark 8.9 concerning the actual existence of Morse germs.

Exercise 8.13 Let $f \in R^{\Delta(n)}$ be defined by the rule

$$\left[(t_1, \ldots, t_n) \mapsto c + u_1 {t_1}^2 + \cdots + u_n {t_n}^2\right],$$

with the u_i's all invertible in R. Then, f is a Morse germ with a non-degenerate singularity at 0.

Solution. The usual rules of derivation give us that the 1-jet of f at 0 is encoded in the $(n+1)$-tuple $(c,0,\dots,0)$. Therefore, $0 \in \Delta(n)$ is a singularity. Moreover, $J^1 f\colon \Delta(n) \longrightarrow R \times R^n$ has the following description:

$$J^1 f(\bar{t}) = \left(c + \sum u_i t_i{}^2,\ 2u_1 t_1 + \cdots + 2u_n t_n\right),$$

which says that 0 is the only possible singularity as the u_i's are invertible. Therefore we only need to check that 0 is non-degenerate or, equivalently, that $\pi \circ J^1 f$ is a submersion at 0. To this end we use Proposition 6.14 (ii) which in this case translates into the true statement that the vectors

$$(u_1,0,\dots,0),(0,u_2,\dots,0),\dots,(0,\dots,0,u_n)$$

are linearly independent.

Remark 8.14 One of the central results in classical Morse theory [2, 47, 54, 11, 87, 91] is the construction of a local chart (for a manifold) or a change of coordinates (for a part of some $\mathbb{R}^n$) around a non-degenerate singularity, making the function 'look like' a non-degenerate quadratic form. In our setting the result can be proven as well. This will be the key to the stability of Morse germs on account of the proposition below.

Proposition 8.15 *Any germ $f \in R^X$ of the form*

$$\left[(t_1,\dots,t_n) \mapsto c + u_1 t_1^2 + \cdots + u_n t_n^2\right]$$

is stable.

Proof By Mather's theorem we need only to check for infinitesimal stability. Without loss of generality we may assume that $f \in R^{\Delta(n)}$. In this case, infinitesimal stability for f means the validity of

$$\forall \omega \in \mathrm{Vect}(f)\, \exists \sigma \in \mathrm{Vect}(R^n)\, \exists \tau \in \mathrm{Vect}(R)\ [\omega = \alpha_f(\sigma) \oplus \beta_f(\tau)].$$

Let ω be a given vector field along f, that is $\omega(x) = (f(x), \underline{\omega}(x))$. We may assume that the principal part at 0 is 0, that is, $\underline{\omega}(0) = 0$. Otherwise, just consider any vector field τ on R such that $\tau(f(0)) = \underline{\omega}(0)$ and then take $\omega - \tau \circ f$. Using Postulate F, it follows from the data $\underline{\omega}\colon \Delta(n) \longrightarrow R$, with $\underline{\omega}(0) = 0$, that there exist germs $h_1,\dots,h_n\colon \Delta(n) \longrightarrow R$ such that $\underline{\omega}(t) = \sum_{i=1}^n h_i(\bar{t}) \cdot t_i$.

The required vector field σ on R^n is that whose principal part is

$$\underline{\sigma}(\bar{t}) = \left(\frac{1}{2}u_1{}^{-1}h_1(\bar{t}),\dots,\frac{1}{2}u_n{}^{-1}h_n(\bar{t})\right).$$

□

Lemma 8.16 *Let $(\mathscr{E}, R)$ be a model of SDG. Assume that $f \in R^{R^n}$ has a second-order Taylor polynomial of the form $\sum_{i=1}^n a_{ij}x_i x_j$. Assume that f has a singularity at 0 and that it is non-degenerate. Then it is possible to find coordinates z_i such that the new Taylor polynomial is $a_1 {z_1}^2 + \cdots + a_n {z_n}^2$.*

Proof The statement of the Lemma is equivalent to the existence of a linear isomorphism φ represented by a non-singular matrix A such that $f \circ \varphi$ has the desired Taylor polynomial.

The usual rules of derivation give us the following equality

$$\frac{\partial (f \circ \varphi)}{\partial x}(a) = \frac{\partial f}{\partial x}(\varphi(a)) \cdot \varphi'(a) = \frac{\partial f}{\partial x}(\varphi(a)) \cdot A$$

and therefore

$$\frac{\partial^2 (f \circ \varphi)}{\partial x^2} = A^t \cdot \frac{\partial^2 (f)}{\partial x^2} \cdot A \, .$$

Therefore, the result will be proved if we show that there is a matrix A which diagonalizes the non-degenerate symmetric bilinear form associated to $\frac{\partial^2 (f \circ \varphi)}{\partial x^2}$, the Hessian of f. This is done just as in the classical setting by using elementary row operations and the corresponding column operations until the matrix is in diagonal form. Multiplying together all non-singular matrices corresponding to the elementary row operations and the transpose matrices corresponding to the elementary column operations one arrives at the desired non-singular matrix. It can be checked that the classical proof is intuitionistically valid for any local ring in a topos, and such is the case for R on account of Axiom K. □

Theorem 8.17 *Every Morse germ $g \in R^{\Delta(n)}$ with a singularity is equivalent to a sum of a quadratic form and a constant.*

Proof The following considerations will bring about simplifications of the proof.

- We may assume that $g(0) = 0$, as a suitable change of coordinates makes no difference for the assertion to be shown.
- We may also assume that the 2-jet of g is of the form $a_1 {x_1}^2 + \cdots + a_n {x_n}^2$ where all the a_i's are invertible in R. This is a consequence of Lemma 8.16.
- With the above reductions, we have that $g = f + \varphi$ where $f = a_1 {x_1}^2 + \cdots + a_n {x_n}^2$ and φ is a germ vanishing on $D_2(n)$. Notice that φ has a zero of order three at 0.

Our goal is to prove that $f \sim g = f + \varphi$ as claimed, and for that the homotopic method is useful. First, join f to g by the path $f + t\varphi$, with $t \in [0,1]$. Next show that it is possible to find a one-parameter family of local diffeomorphisms

$$x + \Delta(n) \mapsto \Phi(t,x) \in \Delta(n)$$

such that

$$\begin{aligned} (f+t\varphi)\Phi(t,x) &= f(x) && \forall x \in \Delta(n)\ \forall t \in [0,1] \\ \Phi(0,x) &= x && \forall x \in \Delta(n) \\ \Phi(t,0) &= 0 && \forall t \in [0,1]. \end{aligned}$$

In this case, $\Phi(1,-)$ will do the job.

A way to obtain the Φ_t for $t \in [0,1]$ in our context is to do so as integral curves for suitable vector fields δ_t or, equivalently, for a compactly supported time-dependent vector field δ—that is, as solutions of

$$\frac{d\Phi}{dt}(x,t) = \delta(\Phi(x,t),t).$$

The equations for δ can be obtained by taking derivatives of

$$(f+t\varphi)\Phi(t,x)$$

with respect to the parameter t. This gives

$$\varphi(\Phi(x,t)) + \frac{d(f+\varphi t)}{dt}\Phi(x,t) \cdot \frac{d\Phi}{dt}(x,t) = 0$$

for each t. If the principal part of δ is expressed by the functions $(\delta_{t_1}, \ldots, \delta_{t_n})$, then we have

$$\varphi_{|\Phi(x,t)} \equiv -\sum_{i=1}^{n} \delta_{t_i} y_{i|\Phi(x,t)}$$

where $y_i = 2a_i x_i + t\varphi_{x_i}$. Therefore $-\varphi \equiv \sum_{i=1}^{n} \delta_{t_i} y_i$, both sides seen as functions of (x,t).

Next we see that, for each $t \in [0,1]$, the determinant

$$\left| \frac{\delta(y)}{\delta(x)}(0,t) \right|$$

is invertible since it equals the determinant

$$\begin{vmatrix} 2a_1 & \cdots & 0 \\ \vdots & \ddots & \vdots \\ 0 & \cdots & 2a_n \end{vmatrix},$$

and Φ_{x_i} vanishes on $D(n)$, as φ did on $D_2(n)$.

It now follows from Postulate I.I, the postulate of infinitesimal inversion

(Postulate 6.17), that for each t, y is a bijection and (y,t) defines a new system of coordinates. In this new system φ takes on the form

$$\varphi(y,t) = \sum_{i=1}^{n} \psi_i(y,t)y_i$$

as φ has no component in t and $\psi_i(0,t) = 0$, $\varphi(0) = 0$. Therefore, $\delta_{t_i} = \psi_i$ as functions on y, and the integral curves give the wanted solution. □

Theorem 8.18 *Let R^X with $X = \neg\neg\{x_0\}$ where $(\mathscr{E},R)$ is a model of SDT. A germ $f \in R^X$ is stable if and only if it is a Morse germ.*

Proof We claim that the property of being a Morse germ is a generic property for the pair $(n,1)$ of dimensions. Firstly, it follows immediately from Proposition 8.15 and Theorem 8.17 that Morse germs are stable. Now, from Proposition 8.10 follows that they are also dense in $R^{\Delta(n)}$ for the weak (hence the intrinsic) topological structure. □

Remark 8.19 In addition to the cases already considered one may obtain in a similar way several other genericity results within a model of SDT. We mention in passing that examples of all stable germs in low dimensions for germs in Y^X that are classically known are, in addition to 1) immersions with normal crossings where $p \geq 2n$, and 2) Morse germs R^X, also 3) Whitney theorem where $p = n = 2$, folds, cusps, 4) for $n = p = 1$ submersions with folds, 5) for $n = p = 3$, folds, cusps, elliptics, and 5) for $n = p = 4$, folds, cusps, elliptic, umbillic. We refer to [47] for a good exposition of these cases. It may be worth remarking that for $p = n = 9$ there are no stable mappings [108].

PART FIVE

SDT AND DIFFERENTIAL TOPOLOGY

Our goals in this book are twofold. The first is to achieve conceptual simplicity by a judicious choice of axioms in the setting of topos theory. The second is to make sure that our results apply to classical mathematics. To this end we revisit the notion of a well adapted model of SDG, extend it to SDT, and then assume the existence of one such. An application of the existence of such a model to the theory of unfoldings is then given.

9
Well Adapted Models of SDT

The construction of several models of SDG begins with the consideration of the *algebraic theory* [70] $\mathscr{C}^\infty$, whose m-tuples of n-ary operations are given by the C^∞-mappings $\mathbb{R}^n \longrightarrow \mathbb{R}^m$ and whose equations are those that are true in general for such smooth mappings. Such a sort of theory was already mentioned by F. W. Lawvere [73] as a tool for getting the standard differential calculus to be amenable to the synthetic method proposed. It was then used by E. Dubuc [32, 34] in the construction of models of SDG. In this chapter we extend the notion of a well adapted model of SDG to one of SDT.

9.1 The Algebraic Theory of C^∞-Rings

Definition 9.1 A C^∞*-ring* A in a category $\mathscr{E}$ with finite products is a model (in $\mathscr{E}$) of the algebraic theory $\mathbb{T}_\infty$ whose n-ary operations are given by

$$\mathbb{T}_\infty(n,1) = C^\infty(\mathbb{R}^n, \mathbb{R})$$

where $\mathbb{R}$ denotes the object of real numbers in *Set*. Equivalently, A is a product preserving functor

$$A\colon \mathscr{C}^\infty \longrightarrow \mathscr{E},$$

where $\mathscr{C}^\infty$ is the category of Euclidean spaces, $\mathbb{R}^n$, $n \geq 0$, and smooth functions. A morphism of C^∞-rings regarded as functors is a natural transformation. Denote by $\mathscr{A}$ the category of C^∞-rings (in *Set*).

Examples of C^∞-rings are $C^\infty(\mathbb{R}^n, \mathbb{R})$ and $C^\infty(N, \mathbb{R})$, with N a C^∞-manifold in the usual sense. In what follows, we will denote the set $C^\infty(\mathbb{R}^n, \mathbb{R})$ by simply $C^\infty(\mathbb{R}^n)$ and similarly $C^\infty(N)$.

We have that $C^\infty(\mathbb{R}^n)$ is the *free* C^∞*-ring* in n generators, on account of the bijection that exists between n-tuples of elements $(a_1,\ldots,a_n) \in A^n$ and C^∞-ring homomorphisms $\varphi \colon C^\infty(\mathbb{R}^n) \longrightarrow A$, the correspondence given by evaluation at the n projections $\pi_i \colon \mathbb{R}^n \longrightarrow \mathbb{R}, i = 1,\ldots,n$.

It is also easy to verify that if A is a C^∞-ring and $I \subset A$ is an *ideal* (in the sense of C^∞-rings), the quotient A/I is again a C^∞-ring.

Definition 9.2 A C^∞-ring is said to be of *finite type* if it is equivalent to one of the form $C^\infty(\mathbb{R}^n)/I$ where $I \subset C^\infty(\mathbb{R}^n)$ is an ideal, and said to be *finitely presented* if it is of finite type defined by a finitely generated ideal. Denote by $\mathscr{A}_{FT} \subset \mathscr{A}$ the full subcategory whose objects are the C^∞-rings of finite type, and similarly $\mathscr{A}_{FP} \subset \mathscr{A}$ for those of finite presentation.

The category $\mathscr{A}$ has finite colimits. Denote by $\otimes_\infty$ the binary coproduct. The initial object is $\mathbb{R}$. For C^∞-rings of finite type (respectively, finitely presented), the binary coproduct is given by

$$C^\infty(\mathbb{R}^n)/I \otimes_\infty C^\infty(\mathbb{R}^m)/J = C^\infty(\mathbb{R}^{n+m})/(I,J)$$

therefore it restricts to the full subcategory $\mathscr{A}_{FT} \subset \mathscr{A}$ of C^∞-rings of finite type (respectively to $\mathscr{A}_{FP} \subset \mathscr{A}$ of finitely presented C^∞-rings). Also, $\mathbb{R} = C^\infty(\mathbb{R}^0) = C^\infty(\mathbf{1})$ belongs to $\mathscr{A}_{FP}$.

In what follows we shall need some basic facts of the theory of C^∞-manifolds. We refer the reader to standard sources such as [68].

Recall that for N an n-dimensional (paracompact) C^∞-manifold, smooth functions $h_1,\ldots,h_k \colon N \longrightarrow \mathbb{R}$ (whose set of zeroes is denoted $Z(h_1,\ldots,h_k)$) are said to be *independent* if for each $p \in Z(h_1,\ldots,h_k)$ the rank of the Jacobian $\left(\left. \frac{\partial h_i}{\partial x_j} \right|_{x=p} \right)$ is equal to k.

Lemma 9.3 *Let N be an n-dimensional (paracompact) C^∞-manifold and let $h_1,\ldots,h_k \colon N \longrightarrow \mathbb{R}$ be independent functions. Then $M = Z(h_1,\ldots,h_k)$ is a C^∞-submanifold of N of dimension $(n-k)$ and the restriction*

$$C^\infty(N) \longrightarrow C^\infty(M)$$

is a quotient in $\mathscr{A}$ with kernel $(h_1,\ldots,h_k)$, that is,

$$C^\infty(M) = C^\infty(N)/(h_1,\ldots,h_k).$$

Proof The proof involves three fundamental results, to wit the **I.F.T.** (Inverse Function Theorem), the **L.H.L.** (Local Hadamard Lemma) and **P.U.** (Partitions of Unity). We sketch the proof below.

(a) By the (**I.F.T.**), for every $p \in M$ there is an open $U \subset N$, $V \subset \mathbb{R}^n$, $p \in U$,

$0 \in V$, and a diffeomorphism $\theta\colon U \approx V$ such that $\theta(p) = 0$ and the diagram

$$\begin{array}{ccc} M \cap U & \xrightarrow{\langle h_1,\dots,h_k\rangle} & \mathbb{R}^k \\ \Big\downarrow{\scriptstyle\theta} & & \Big\downarrow{\scriptstyle=} \\ \mathbb{R}^{(n-k)} \cap V & \xrightarrow[(x_{n-k+1},\dots,x_n)]{} & \mathbb{R}^k \end{array}$$

commutes. This defines the structure of a closed manifold of N on M and for each $p \in N$,

$$C_p^\infty(N) \twoheadrightarrow C_p^\infty(M)$$

is locally surjective, where $C_p^\infty(N)$ (respectively $C_p^\infty(M)$) denote the rings of germs at p of the indicated smooth mappings.

(b) We now claim that the kernel of the above morphism is

$$(h_1|_p, \dots, h_k|_p).$$

The proof, which follows, is an application of the Local Hadamard's Lemma (**L.H.L.**). If $V = V_1 \times \cdots \times V_n \subset \mathbb{R}^n$ is a product of open intervals of $\mathbb{R}$ and $\varphi\colon V \longrightarrow \mathbb{R}$ is any (smooth) C^∞-function, then there exist unique C^∞-functions $\psi_i\colon V \times V \longrightarrow \mathbb{R}$ such that

$$\varphi(\bar{x}) - \varphi(\bar{y}) = \sum_{i=1}^{n} (x_i - y_i) \cdot \psi_i(\bar{x}, \bar{y})$$

for any $\bar{x}, \bar{y} \in V$. Now, for φ such that

$$\varphi(x_1, \dots, x_{n-k}, 0, \dots, 0) = 0$$

there exist unique $\psi_i\colon V \longrightarrow \mathbb{R}$ such that

$$\varphi(\bar{x}) = \sum_{i=n-k+1}^{n} x_i \cdot \psi_i(\bar{x})$$

taking $\bar{x} = (x_1, \dots, x_n)$ and $\bar{y} = (x_1, \dots, x_{n-k}, 0, \dots, 0)$.

Let $f\colon U \longrightarrow \mathbb{R}$, $p \in U$ such that $f|_{M\cap U} = 0$. Setting

$$\varphi = f \circ \theta^{-1}\colon V \longrightarrow \mathbb{R}$$

gives $\varphi(x_1, \dots, x_{n-k}, 0, \dots, 0) = 0$, since

$$\theta^{-1}(x_1, \dots, x_{n-k}, 0, \dots, 0) \in M \cap U$$

for $(x_1, \dots, x_{n-k}, 0, \dots, 0) \in V$.

We get

$$(f \circ \theta^{-1})(\bar{x}) = \sum_{i=n-k+1}^{n} x_i \cdot \psi_i(\bar{x})$$

for $\bar{x} \in V$, so

$$f(\bar{x}) = \sum_{i=1}^{k} h_i(\bar{x}) \cdot \psi_{n-k+1}(\theta(\bar{x}))$$

for $\bar{x} \in U$, so that $f \in (h_1|_p, \ldots, h_k|_p)$. Conversely, if $f \in (h_1|_p, \ldots, h_k|_p)$ then $f|_{M \cap U} = 0$.

(c) We now use **P.U.** to globalize the data. Let us recall what it says. For M a paracompact C^∞-manifold, if $\{U_\alpha \mid \alpha \in I\}$ is any open covering, there is a locally finite refinement $W_\beta \subset U_{\alpha_\beta}$. This means that for any $x \in M$ there is V such that $x \in V$ and such that $\{\beta \mid V \cap W_\beta \neq \emptyset\}$ is finite. Associated to this $\{W_\beta\}$ there is a partition of unity, that is, C^∞-functions $\varphi_\beta : M \longrightarrow \mathbb{R}$ such that

$$\text{supp}(\varphi_\beta) = \overline{\{x \in M \mid \varphi_\beta(x) \neq 0\}} \subset W_\beta$$

and such that $\sum \varphi_\beta = 1$.

We wish to show that the map $C^\infty(N) \longrightarrow C^\infty(M)$ is surjective with kernel $(h_1, \ldots, h_k)$ so that $C^\infty(M) \approx C^\infty(N)/(h_1, \ldots, h_k)$. Let $h \in C^\infty(M)$ be such that $h|_M = 0$. We wish to show that $h \in (h_1, \ldots, h_k)$. By the local version, there are open sets $\{U_\alpha \subset N \mid \alpha \in \Gamma\}$ covering M such that

$$h|_{U_\alpha} \in (h_1|_{U_\alpha}, \ldots, h_k|_{U_\alpha}).$$

Let $U_i = \{x \mid h_i(x) \neq 0\}$. Then $\{U_\alpha \mid \alpha \in \Gamma \cup \{1, \ldots, k\}\}$ is an open cover of N and the above still holds for all $\alpha \in \Gamma \cup \{1, \ldots, k\}$. Get $\{W_\beta\}$ a locally finite refinement of $\{U_\alpha\}$ and associated partition of unity $\{\varphi_\beta\}$. By the above, for each β there exists $g_i^\beta : W_\beta \longrightarrow \mathbb{R}$ for $i = 1, \ldots, k$, such that

$$h|_{W_\beta} = \sum_{i=1}^{k} g_i^\beta \cdot h_i|_{W_\beta}.$$

$$\varphi_\beta \cdot h = \sum_{i=1}^{k} \varphi_\beta \cdot g_i^\beta \cdot h_i$$

and so

$$h = \Big(\sum_\beta \varphi_\beta\Big) \cdot h = \sum_\beta (\varphi_\beta \cdot h) = \sum_\beta \Big(\sum_{i=1}^k \varphi_\beta \cdot g_i^\beta \cdot h_i\Big)$$
$$= \sum_{i=1}^k \Big(\sum_\beta \varphi_\beta \cdot g_i^\beta\Big) \cdot h_i = \sum_{i=1}^k g_i \cdot h_i$$

where g_i denotes $\sum_\beta (\varphi_\beta \cdot g_i^\beta)$. In conclusion, $h \in (h_1, \ldots, h_k)$.

□

Let $\mathscr{C}$ be the category whose objects are the *opens of Euclidean spaces* and whose morphisms are smooth mappings. Consider the functor

$$C^\infty(-)\colon \mathscr{C} \longrightarrow \mathscr{A}^{\mathrm{op}}$$

where $U \mapsto C^\infty(U)$ and

$$U \subset V \mapsto C^\infty(V) \longrightarrow C^\infty(U)$$

is given by restriction.

Proposition 9.4 *The functor $C^\infty(-)\colon \mathscr{C} \longrightarrow \mathscr{A}^{\mathrm{op}}$ preserves (a) open inclusions, (b) finite products, (c) equalizers of independent functions, and it factors through ${\mathscr{A}_{FP}}^{\mathrm{op}} \subset \mathscr{A}^{\mathrm{op}}$.*

Proof In what follows we make use of Lemma 9.3 without mention.

(a) It is enough to prove the statement for open inclusions of the form $U \subset \mathbb{R}^n$. Let φ be a smooth characteristic map of U, that is $U = \varphi^{-1}(\mathbb{R}^*)$, for $\mathbb{R}^*$ denoting the subset of invertible (that is, non-zero) elements of $\mathbb{R}$.

The map $\gamma\colon U \longrightarrow \mathbb{R}^{n+1}$ defined by $\bar{x} \mapsto (\bar{x}, 1/\varphi(\bar{x}))$ is injective and identifies $U \approx Z(1 - \varphi(\bar{x}) \cdot y)$. Notice that $1 - \varphi(\bar{x}) \cdot y\colon \mathbb{R}^{n+1} \longrightarrow \mathbb{R}$ is independent as, for $(\bar{x}, y)$ such that $\varphi(\bar{x}) \cdot y = 1$, the Jacobian

$$\left(\frac{\partial \varphi}{\partial x_1} \cdot y, \ldots, \frac{\partial \varphi}{\partial x_n} \cdot y, \varphi(\bar{x})\right)$$

has rank 1 since $\varphi(\bar{x})$ is invertible. Therefore,

$$C^\infty(\mathbb{R}^{n+1})/(1 - \varphi(\bar{x})y) \approx C^\infty(U)$$

and the restriction

$$C^\infty(\mathbb{R}^n) \longrightarrow C^\infty(U)$$

is surjective. This says that $C^\infty(U)$ is finitely presented, hence the desired factorization. To see that also $C^\infty(\mathbb{R}^n) \longrightarrow C^\infty(U)$ is surjective, let A be any C^∞-ring and $h, h'\colon C^\infty(U) \longrightarrow A$ any two C^∞-ring homomorphisms which agree on $C^\infty(\mathbb{R}^n)$. This means that they agree on the images $\pi_i|_U$ of the n projections, but also on $\varphi(\bar{x}) \in C^\infty(\mathbb{R}^n)$ as well as on $1/\varphi(\bar{x})$ for every $\bar{x} \in U$.

(b) That binary products (hence finite products) are preserved is proved using that whereas

$$C^\infty(U) \otimes_\infty C^\infty(V) = C^\infty(\mathbb{R}^{n+1+m+1})/(1-\varphi(\bar{x})\cdot y, 1-\psi(\bar{u})\cdot v),$$

we have

$$U \times V = Z(1-\varphi(\bar{x})\cdot y, 1-\psi(\bar{u})\cdot v)$$

and the two functions of $\bar{x}, y, \bar{u}, v$ are independent as consideration of the corresponding Jacobian shows. It follows that

$$C^\infty(U \times V) \approx C^\infty(U) \otimes_\infty C^\infty(V).$$

(c) Let

$$E \hookrightarrow U \underset{f_2}{\overset{f_1}{\rightrightarrows}} V$$

be an equalizer in $\mathscr{C}$ of independent functions f_1, f_2. Now, $E = Z(f-g)$ and $f-g$ is independent : indeed, if the Jacobian

$$\begin{pmatrix} \frac{\partial f}{\partial x_1} & \cdots & \frac{\partial f}{\partial x_n} \\ \\ \frac{\partial g}{\partial x_1} & \cdots & \frac{\partial g}{\partial x_n} \end{pmatrix}$$

has rank 2, then the matrix

$$\left(\frac{\partial f}{\partial x_1} - \frac{\partial g}{\partial x_1} \cdots \frac{\partial f}{\partial x_n} - \frac{\partial g}{\partial x_n} \right)$$

has rank 1. Therefore,

$$C^\infty(E) \approx C^\infty(U)/(f_1 - f_2).$$

□

The category $\mathscr{M}^\infty$ of all (paracompact) C^∞-manifolds and smooth mappings does not have arbitrary pullbacks but it has transversal pullbacks. Recall Definition 9.9 of when a pullback diagram

$$\begin{array}{ccc} S & \xrightarrow{k} & M_1 \\ {\scriptstyle h}\downarrow & \lrcorner & \downarrow{\scriptstyle f} \\ M_2 & \xrightarrow[g]{} & N \end{array}$$

is said to be *transversal*. This is the case if for all $p \in S, x = k(p)$ and $y = h(p)$, the image under df_x of $T_x(M_1)$ and the image of $T_y(M_2)$ under dg_y generate $T_z(N)$, where $z = f(x) = g(y)$.

We now recall without proof the following easily established result

Lemma 9.5 *[38] For $\mathscr{E}$ a category with finite limits, a functor $F : \mathscr{C} \longrightarrow \mathscr{E}$ preserves transversal pullbacks and* 1 *if and only if it preserves open inclusions, finite products, and equalizers of independent functions. This equivalence extends to natural transformations and homomorphisms of C^∞-rings.*

Corollary 9.6 *The functor $C^\infty : \mathscr{C} \longrightarrow \mathscr{A}^{\mathrm{op}}$ factors through $\mathscr{A}_{FP}^{\mathrm{op}} \hookrightarrow \mathscr{A}^{\mathrm{op}}$ and preserves transversal pullbacks and* 1.

The following is a geometric characterization of C^∞-rings.

Theorem 9.7 *[24] Let $\mathscr{E}$ be a category with finite limits. Then there is a bijection between C^∞-rings in $\mathscr{E}$ and functors $F : \mathscr{C} \longrightarrow \mathscr{E}$ preserving transversal pullbacks and the terminal object* 1.

Proof By the theory of algebraic theories, a C^∞-ring A is determined by the value of $F : \mathscr{A}_{FP}{}^{\mathrm{op}} \longrightarrow \mathscr{E}$ at the free C^∞-ring on one generator, which is $C^\infty(\mathbb{R})$. To A corresponds $\mathrm{Spec}_A : \mathscr{A}_{FP}{}^{\mathrm{op}} \longrightarrow \mathscr{E}$. Given A take the composite

$$\mathscr{C} \xrightarrow{C^\infty(-)} \mathscr{A}_{FP}{}^{\mathrm{op}} \xrightarrow{\mathrm{Spec}_A} \mathscr{E}.$$

It preserves transversal pullbacks and 1.

Conversely, given any $F : \mathscr{C} \longrightarrow \mathscr{E}$ preserving transversal pullbacks and 1, evaluate F at $\mathbb{R}$. The correspondence $F \mapsto F(\mathbb{R})$ has values in the category of $\mathscr{C}^\infty$-rings in $\mathscr{E}$. Moreover, F is totally determined by its value at $\mathbb{R}$. This is because F preserves transversal pullbacks and 1, and since the pullbacks

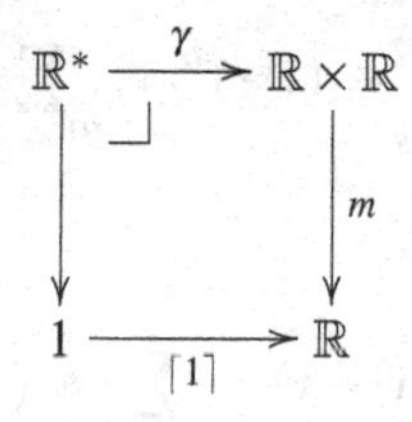

and

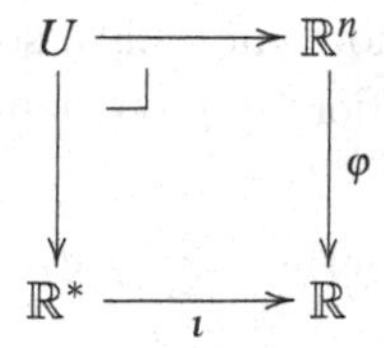

where $\gamma(x) = (x, \frac{1}{x})$, m is multiplication, φ is a characteristic map of U and ι is the inclusion, are both transversal.

To finish the proof about the correspondence on objects, we just note that $\mathrm{Spec}_A(C^\infty(\mathbb{R})) = A$. We leave it to the reader to extend this correspondence to one of morphisms. □

In the proof of Proposition 9.4 it was shown that for any open subset $U \subset \mathbb{R}^n$, $C^\infty(U)$ is a finitely presented C^∞-ring. Another important class of examples of objects of $\mathscr{A}_{FP}$ is given by the Weil algebras, as shown next.

Proposition 9.8 *Any Weil algebra W over $\mathbb{R}$ has a canonical structure of a (finitely presented) C^∞-ring such that for any C^∞-ring C,*

$$\hom_{\mathbb{R}-\text{algebras}}(W,C) = \hom_{C^\infty-\text{rings}}(W,C).$$

In addition, for any $B \in \mathscr{A}$, there is a C^∞-ring structure on $B \otimes_{\mathbb{R}} W$ such that the canonical $\mathbb{R}$-algebra map $j\colon B \longrightarrow B \otimes_{\mathbb{R}} W$ is a morphism of C^∞-rings. In case the $\mathbb{R}$-coproduct is a C^∞-ring, then $\otimes_{\mathbb{R}}$ agrees with $\otimes_\infty$.

Proof Let $W = \mathbb{R} \oplus I$ be such that $I^{k+1} = 0$. Then

$$\begin{aligned} B \otimes_{\mathbb{R}} W &= B \otimes_{\mathbb{R}} (\mathbb{R} \oplus I) \\ &\cong (B \otimes_{\mathbb{R}} \mathbb{R}) \oplus (B \otimes_{\mathbb{R}} I) \\ &\cong j(B) \oplus (B \otimes_{\mathbb{R}} I). \end{aligned}$$

Every $r \in B \otimes_{\mathbb{R}} W$ is then of the form $r = j(x) + y$ for $x \in B, y^{k+1} = 0$. To give a C^∞-ring structure we need to define $\varphi(\bar{x} + \bar{y})$ for $\varphi \in C^\infty(\mathbb{R}^n)$ and $r_i = j(x_i) + y_i$.

By Hadamard's Lemma, and for the given $k \geq 0$, there exist unique smooth mappings $\varphi_\alpha\colon \mathbb{R}^n \longrightarrow \mathbb{R}$, $\psi_\beta\colon \mathbb{R}^n \times \mathbb{R}^n \longrightarrow \mathbb{R}$, such that for all $\bar{x}, \bar{y} \in \mathbb{R}^n$, one has

$$\varphi(\bar{x} + \bar{y}) = \sum_{|\alpha| \leq k} \varphi_\alpha(\bar{x}) \cdot \bar{y}^\alpha + \sum_{|\beta| = k+1} \psi_\beta(\bar{x}, \bar{y}) \cdot \bar{y}^\beta .$$

This forces

$$\varphi(j(\bar{x}) + y) = \sum_{|\alpha| \leq k} \varphi_\alpha(j(\bar{x}))$$

since $\bar{y}^{k+1} = 0$ and j is a C^∞-homomorphism so that $\varphi(j(\bar{x})) = j(\varphi_\alpha(\bar{x}))$ and so, the way to interpret the action of φ on elements of $B \otimes_{\mathbb{R}} W$ is none other than

$$\varphi(j(\bar{x} + \bar{y})) = \sum_{|\alpha| \leq k} j(\varphi_\alpha(\bar{x})) \cdot \bar{y}^\alpha .$$

In particular, for $W = \mathbb{R} \otimes_{\mathbb{R}} W$ we get, for $r_i = j(x_i) + y_i, x_i \in \mathbb{R}, y_i \in I$, $\varphi(r_i) = \varphi(j(x_i) + y_i) = \sum_{|\alpha| \leq k} \varphi_\alpha(j(x_i)) \cdot \bar{y}^\alpha$ is a polynomial, hence the agreement between $\mathbb{R}$-algebra maps and C^∞-homomorphisms. □

9.2 The Theory of Well Adapted Models of SDT

Having laid down the axioms and postulates of SDT, we now turn to the sort of well adapted models of SDT that satisfy the additional axioms that constitute SDT.

Denote by $\mathscr{M}^\infty$ the category of (paracompact, finite dimensional) smooth manifolds and smooth mappings. A definition of the notion of a manifold M is to regard it as a collection of open sets $\{U_\alpha \subset \mathbb{R}^n \mid \alpha \in I\}$ together with patching data, that is, to regard M as a quotient of the disjoint union of the open sets U_α. We refer the reader to [68] for further details.

The category $\mathscr{M}^\infty$ of all (paracompact) C^∞-manifolds and smooth mappings does not have arbitrary pullbacks but it has transversal pullbacks. We recall here the definition.

Definition 9.9 In the category $\mathscr{M}^\infty$, a pullback diagram

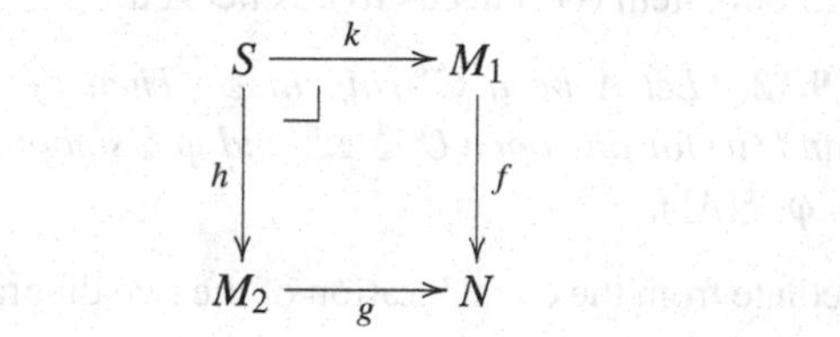

$$\begin{array}{ccc} S & \xrightarrow{k} & M_1 \\ {\scriptstyle h}\downarrow & \lrcorner & \downarrow{\scriptstyle f} \\ M_2 & \xrightarrow[g]{} & N \end{array}$$

is said to be *transversal* if for all $p \in S$, $x = k(p)$ and $y = h(p)$, the image under f of $T_x(M_1)$ and the image of $T_y(M_2)$ in N generate $T_z(N)$, where $z = f(x) = g(y)$.

Definition 9.10 A *well adapted model of SDT* is a well adapted model $(\mathscr{E}, R)$ of SDG which in addition satisfies Axiom **G** (germs representability), Axiom **M** (the germs representing objects are tiny), Postulate **E** (covering property of the Euclidean topological structure), Postulate **S** (existence and uniqueness of solutions to parametrized ordinary differential equations), Postulate **I.I** (infinitesimal inversion) and Postulate **D** (density of regular values).

In this section we are interested in well adapted models $(\mathscr{E}, R)$ constructed using C^∞-rings. A C^∞-ring A in $\mathscr{E}$ will be here identified (even notationally) with a functor

$$A \colon \mathscr{C} \longrightarrow \mathscr{E}$$

where A preserves transversal pullbacks and 1 (cf. Theorem 9.7).

Without any additional axioms, a C^∞-ring A possesses an order $>$, defining

$$A_{>0} = A(\mathbb{R}_{>0}) \hookrightarrow A(\mathbb{R}) = A.$$

In particular, the order $>$ on A is strict ($\neg(0 > 0)$) and it is compatible with the ring operations—the latter by functoriality of A and the fact that, since polynomials are smooth functions, every C^∞-ring is a ring. Similarly, let $A_{<0} = A(\mathbb{R}_{<0})$. Since $\mathbb{R}_{<0} = \{x \in \mathbb{R} \mid -x \in \mathbb{R}_{>0}\}$, we have, for $x \in A$ that $x \in A_{<0}$ if and only if $-x \in A_{>0}$.

In the following propositions, for

$$A\colon \mathscr{C} \longrightarrow \mathscr{E},$$

the expression 'the functor A preserves a certain open covering' is meant to be interpreted as 'the functor A sends the given open covering in $\mathscr{C}$ to a jointly epimorphic family in $\mathscr{E}$'.

Proposition 9.11 *A C^∞-ring A in $\mathscr{E}$ is an ordered ring if and only if the functor $A\colon \mathscr{C} \longrightarrow \mathscr{E}$ preserves the open covering*

$$\mathbb{R}^* = \mathbb{R}_{>0} \cup \mathbb{R}_{<0}.$$

Proof In 2.13 only item (O3) needs to be checked. □

Proposition 9.12 *Let A be a C^∞-ring in $\mathscr{E}$. Then the following hold: (i) $A(\mathbb{R}^*) = A^*$ and (ii) for any open $U \subset \mathbb{R}^n$ and φ a smooth characteristic map of U, $A(U) = \varphi^{-1}(A^*)$.*

Proof Immediate from the consideration of the two diagrams in Theorem 9.7. □

Proposition 9.13 *A C^∞-ring A in $\mathscr{E}$ is local as a ring if and only if the functor $A\colon \mathscr{C} \longrightarrow \mathscr{E}$ preserves the open covering*

$$\mathbb{R} = \mathbb{R}^* \cup (1 - \mathbb{R}^*).$$

Proof As already observed, A possesses a strict order, $0 = 1$ leads to contradiction as $1 > 0$. This takes care of condition (i). Hence only (ii) needs to be considered and the assertion is then obvious. □

The following statement is now obvious.

Proposition 9.14 *A commutative ring A in a topos $\mathscr{E}$ with a natural numbers object is Archimedean if and only if $A\colon \mathscr{C} \longrightarrow \mathscr{E}$ preserves the open covering*

$$\mathbb{R} = \bigcup_{n \in \mathbb{N}} (-n, n).$$

The following lemma is shown in [24] and was obtained in collaboration with A. Joyal.

Lemma 9.15 *A functor $A\colon \mathscr{C} \longrightarrow \mathscr{E}$ where $\mathscr{E}$ is a topos with a natural numbers object $\mathbb{N}$ preserves all open coverings in $\mathscr{C}$ if and only if it preserves open coverings of the form*

$$\mathbb{R} = \mathbb{R}^* \cup (1 - \mathbb{R}^*)$$

and

$$\mathbb{R} = \bigcup_{n \in \mathbb{N}} (-n, n) .$$

We now state a geometric characterization of Archimedean local C^∞-rings in a Grothendieck topos $\mathscr{E}$.

Theorem 9.16 *[24] Let $\mathscr{E}$ be a Grothendieck topos. Then there is a bijection between functors $F\colon \mathscr{C} \longrightarrow \mathscr{E}$ preserving transversal pullbacks and 1 and open coverings and Archimedean local C^∞-rings in $\mathscr{E}$.*

Proof The proof follows directly from Proposition 9.11, Proposition 9.13, Proposition 9.14, and Lemma 9.15. □

The connection with well adapted models of ringed toposes in the sense of Definition 2.16 is given next. It relies on the crucial observation that, for $\mathscr{M}^\infty$ the category of (paracompact) C^∞-manifolds, and $\mathscr{C}$ the category of open subsets of objects of the form $\mathbb{R}^n$ and inclusion maps, the inclusion

$$\mathscr{C} \hookrightarrow \mathscr{M}^\infty$$

is *dense* in the sense that, for any $M \in \mathscr{M}^\infty$, there is an open cover $\{U_\alpha \subset M\}_{\alpha \in I}$ with $U_\alpha \approx V_\alpha \in \mathscr{C}$. In other words, what will be used about manifolds is their very definition, that is as locally open subsets of some $\mathbb{R}^n$. More precisely we have the following.

Theorem 9.17 *[24] There is a bijection between (i) well adapted models $(\mathscr{E}, R)$ of ringed toposes in the sense of Definition 2.16 and (ii) Archimedean local C^∞-rings A in $\mathscr{E}$.*

Proof $(i) \Rightarrow (ii)$. Restrict the given functor $\iota\colon \mathscr{M}^\infty \hookrightarrow \mathscr{E}$ to the subcategory $\mathscr{C} \hookrightarrow \mathscr{M}^\infty$ and use Theorem 9.7.

$(ii) \Rightarrow (i)$. Let $A\colon \mathscr{C} \longrightarrow \mathscr{E}$ be an Archimedean local ring in $\mathscr{E}$ viewed as a functor. Let $M \in \mathscr{M}^\infty$. The functor A determines a collection of objects in $\mathscr{E}$

$$\{A(U_\alpha) \longrightarrow A^n\}_{\alpha \in I}$$

together with patching data. Since $\mathscr{E}$ is a topos with a natural numbers object and all limits and colimits, the same construction can be performed in $\mathscr{E}$ thus determining an object $A(M)$ of $\mathscr{E}$.

□

Proposition 9.18 *Let $(\mathscr{E}, R)$ be a well adapted model of ringed toposes constructed using C^∞-rings, with inclusion $\iota\colon \mathscr{M}^\infty \hookrightarrow \mathscr{E}$ sending $\mathbb{R}$ to R. Let W be a Weil algebra. If $\mathscr{E} = \widetilde{\mathscr{B}^{\mathrm{op}}}$ where $\mathscr{B} \hookrightarrow \mathscr{A}_{FP}$ is a full subcategory closed under finite coproducts with Weil algebras, then the object $\iota W = a(\overline{W})$ of $\mathscr{E}$ is tiny in the sense of Definition 1.24.*

Proof Since $W \in \mathscr{B}$, the representable functor $\overline{W} \in Set^{\mathscr{B}}$ is an atom. It follows now from Theorem 1.32 that for $a\colon Set^{\mathscr{B}} \longrightarrow \widetilde{\mathscr{B}^{\mathrm{op}}}$ the associated sheaf functor, $a(\overline{W})$ is an atom. In addition, the Weil algebra W has a (unique) point $W \longrightarrow \mathbb{R}$. In particular, $\iota W \longrightarrow i(\mathbb{R})$ is a (split) epimorphism. □

Definition 9.19 Let A be a C^∞-ring. We call *point* of A any C^∞-ring homomorphism $p\colon A \longrightarrow \mathbb{R}$. The C^∞-ring A is said to be *point determined* if for any $a \in A$ and point $p\colon A \longrightarrow \mathbb{R}$, if $p(a) = 0$, then $a = 0$.

Remark 9.20 (i) Clearly, any C^∞-ring of the form $C^\infty(M)$ for $M \in \mathscr{M}^\infty$ is point determined. Indeed, by 'Milnor's exercise' [88, Problem 1C, p. 11], a point of $C^\infty(M)$ 'is' a point $p \in M$.

(ii) Not every finitely presented C^∞-ring is point determined. Consider for instance a Weil algebra W regarded as a C^∞-ring. It was shown in Proposition 9.8 that any such W is finitely presented. A Weil algebra $W = \mathbb{R}[X_1, \ldots, X_n]/J$ has a unique point $\pi\colon W \longrightarrow \mathbb{R}$ corresponding to $0 \in W$, but W is not determined by its unique point since for a non-zero polynomial f, the condition $f(0) = 0$ does not imply that $f = 0$. Therefore, to restrict to point determined C^∞-rings would exclude the interesting examples on account of the failure of the KL-axiom.

Let $M \in \mathscr{M}^\infty$ and $p \in M$. The C^∞-ring of germs at p of $\mathbb{R}$-valued smooth mappings on M is classically constructed as the quotient

$$C^\infty_p(M) = C^\infty(M)/J(p),$$

with $J(p)$ the ideal of smooth functions $f\colon M \longrightarrow \mathbb{R}$ for which there exists an open subset $U \subset M$ such that $p \in U$ and $f|_U = 0$. An alternative notation for $J(p)$ is $\mathfrak{m}^g_p$. That this is a correct description follows from a smooth version of Tietze's extension theorem and makes $C^\infty_p(M)$ into a C^∞-ring.

The ring $C^\infty_p(M)$ of germs at p of smooth mappings $M \longrightarrow \mathbb{R}$ is not finitely presented as a C^∞-ring. Since important examples of C^∞-rings, such as $C^\infty(U)$

for $U \subset \mathbb{R}^n$ an open subset, are finitely presented, an enlargement of the full subcategory $\mathscr{A}_{FP} \subset \mathscr{A}$ is desirable.

Definition 9.21 An ideal I of $C^\infty(\mathbb{R}^n)$ is said to be *germ determined* (or *local*) if for each $f \in C^\infty(\mathbb{R}^n)$,

$$f \in I \Leftrightarrow \forall x \in \mathbb{R}^n \, (f_x \in I_x)$$

where f_x is the germ of f at x and I_x is the ideal generated by $\{g_x \mid g \in I\}$.

Denote by $\mathscr{B}$ the full subcategory of $\mathscr{A}$ whose objects are those C^∞-rings of finite type of the form $B = C^\infty(\mathbb{R}^n)/J$ where $J \subset C^\infty(\mathbb{R}^n)$ is a germ determined ideal.

Proposition 9.22 *We have inclusions*

$$\mathscr{A}_{FP} \subset \mathscr{B} \subset \mathscr{A}_{FT} \subset \mathscr{A}.$$

Proof For the inclusion $\mathscr{A}_{FP} \subset \mathscr{B}$ we show, more generally, that if $J \subset C^\infty(M)$ is a germ determined ideal and $h \in C^\infty(M)$, then the ideal $I = (J, h)$ is germ determined. In particular, since $\{0\}$ is germ determined, it follows from it that any finitely generated ideal is germ determined.

Let $I = (J, h)$ with J germ determined. Assume that for $f \in C^\infty(M)$, $f_p \in I_p$ for all $p \in M$. This means that the germ of f at p equals the germ at p of some $g^{(p)} \in I$ so that there is some $\gamma^{(p)}$ vanishing on an open $V^{(p)}$ containing p, such that $f - g^{(p)} = \gamma^{(p)}$. Let $\{U_\alpha\}$ be a locally finite refinement of $\{V^{(p)}\}$ with subordinated partition of unity $\{\varphi_\alpha\}$. We then have

$$\varphi_\alpha \cdot f = \varphi_\alpha \cdot g^{(p)} + \varphi_\alpha \cdot \gamma^{(p)} = \varphi_\alpha \cdot g^{(p)}$$

since $\text{supp}(\varphi_\alpha) \subset U_\alpha \subset V^{(p)}$ and $\gamma^{(p)}$ vanishes there. Now, since $g^{(p)} \in I$, also $\varphi_\alpha \cdot f \in I$ for each α. Therefore,

$$\varphi_\alpha \cdot f = j_\alpha + k_\alpha \cdot h$$

with $j_\alpha \in J$, $k_\alpha \in C^\infty(M)$.

Since $\text{supp}(\varphi_\alpha) \subset U_\alpha$, consider the covering $\{U_\alpha, \neg(\text{supp}(\varphi_\alpha))\}$ and associated partition of unity $\{\rho_\alpha, 1 - \rho_\alpha\}$. Since $\rho_\alpha \equiv 1$ on $\text{supp}(\varphi_\alpha)$, we have $\rho_\alpha \cdot \varphi_\alpha = \varphi_\alpha$. Therefore,

$$\begin{aligned} \varphi_\alpha \cdot f &= \rho_\alpha \cdot \varphi_\alpha \cdot f \\ &= \rho_\alpha \cdot j_\alpha + \rho_\alpha \cdot k_\alpha \cdot h \\ &= \bar{j}_\alpha + \bar{k}_\alpha \cdot h \end{aligned}$$

where both $\bar{j}_\alpha$ and $\bar{k}_\alpha$ have support included in U_α. It follows that

$$f = \Big(\sum_\alpha \varphi_\alpha\Big) \cdot f = \sum_\alpha (\varphi_\alpha \cdot f)$$
$$= \sum_\alpha (\bar{j}_\alpha + \bar{k}_\alpha \cdot h) = \sum_\alpha \bar{j}_\alpha + \sum_\alpha (\bar{k_\alpha} \cdot h).$$

The germ at p of $\sum_\alpha \bar{j}_\alpha$ is a finite sum of germs, therefore in J_p. Since J is germ determined $\sum_\alpha \bar{j}_\alpha \in J$. Also, $(\sum_\alpha \bar{k}_\alpha) \cdot h \in (h)$. Hence $f \in I = (J, h)$.

□

An alternative (classical) description of the C^∞-ring $C_p^\infty(M)$ of germs at p of smooth mappings $M \longrightarrow \mathbb{R}$ is useful in what follows. It involves inverting elements in $\mathscr{A}$.

For $B \in \mathscr{A}$ and $\Sigma \subset B$, denote by $B\{\Sigma\}^{-1}$ the universal solution to inverting the elements of Σ in B, if such exists. It can be obtained as a directed colimit of such construction for finite subsets $\Sigma \subset B$ and, in turn, for one element subsets $\Sigma = \{b\}$ for $b \in B$. For rings ($\mathbb{R}$-algebras),

$$B\{b^{-1}\} = B[t]/(t \cdot b - 1)$$

satisfies the desired property. Note now that in this case

$$B[t] = \mathbb{R}[t] \otimes_{\mathbb{R}} B.$$

This construction can be repeated in $\mathscr{A}_{FT}$ by letting, for a C^∞-ring $B = C^\infty(\mathbb{R}^n)/I$,

$$B[t] = C^\infty(\mathbb{R}) \otimes_\infty B \cong C^\infty(\mathbb{R}^{n+1})/I$$

and then letting

$$B\{b^{-1}\} = C^\infty(\mathbb{R}^{n+1})/(\pi_1{}^n I,\ t \cdot b - 1).$$

This construction has the desired universal property in $\mathscr{A}$. Furthermore, if $B \in \mathscr{A}_{FP}$ then so is $B\{b^{-1}\}$.

Theorem 9.23 *For $p \in M$, $M \in \mathscr{M}^\infty$, let*

$$\Sigma_p = \{f \in C^\infty(M) \mid f(p \neq 0)\}.$$

Then there is an isomorphism

$$C^\infty(M)/J(p) \cong C^\infty(M)\{\Sigma_p{}^{-1}\}$$

of C^∞-rings.

Proof ($\Rightarrow$) Show that if $f \in J(p)$ then it becomes 0 in $C^\infty(M)\{\Sigma_p{}^{-1}\}$. Let $f|_U \equiv 0$. To this end, consider $\{U, \neg p\}$ as an open covering, with φ, ψ a subordinated partition of unity. Since $p \in M$ and $\text{supp}(\psi) \subset \neg\{p\}$, $\psi(p) = 0$. Therefore, $\varphi(p) = 1$ and so $\varphi \in \Sigma_p$. Since $\text{supp}(\varphi) \subset U$ and $f|_U \equiv 0$, $\varphi \cdot f$ is defined everywhere and $\varphi \cdot f \equiv 0$. Since $\varphi \in \Sigma_p$, φ is invertible in $C^\infty(M)\{\Sigma_p{}^{-1}\}$, we must have $f = 0$ in it.

($\Leftarrow$) Show that if $g \in \Sigma_p$ then g becomes invertible in $C^\infty(M)/J(p)$. Let $U \subset M$ open such that $p \in U$ and $g|_U \neq 0$. As open covering consider $\{U, \neg\{p\}\}$ with φ, ψ a subordinated partition of unity. We now claim that the product $\varphi \cdot g^{-1}$ is everywhere defined. If $x \in \text{supp}(\varphi) \subset U$, $g(x)^{-1}$ exists and if $x \notin \text{supp}(\varphi)$ then $\varphi(x) = 0$. Notice now that

$$g \cdot (\varphi \cdot g^{-1}) - 1 = \varphi - 1 = -\psi.$$

Therefore it suffices to show that $-\psi \in J(p)$ to prove our claim.

Since $\text{supp}(\psi) \subset \neg\{p\}$, $\psi(p) = 0$ on $\neg\text{supp}(\psi)$ and so $\psi \in J(p)$. □

For B a C^∞-ring and $p\colon B \longrightarrow \mathbb{R}$ a point, let B_p be the C^∞-ring $B\{\Sigma_p^{-1}\}$ (the germ algebra of B at p) where $\Sigma_p = \{f \in B \mid p(f) \neq 0\}$. The homomorphism $B \longrightarrow B_p$ is said to send $f \in B$ to the germ f_p of f at p. For an ideal $I \subset B$, let I_p be the smallest ideal containing $\{f_p \mid f \in I\}$. For any ideal $I \subset B$, the *germ radical* is defined as

$$\hat{I} = \{f \in B \mid \forall p \in B\, (f_p \in I_p)\}.$$

An ideal I of B is germ determined if and only if $I = \hat{I}$. Call B itself germ determined if $\widehat{\{0\}} = \{0\}$. A C^∞-ring B maps to a germ determined C^∞-ring $B/\widehat{\{0\}}$ with the same points.

It follows that the inclusion $\mathscr{B} \hookrightarrow \mathscr{A}_{FT}$ has a right adjoint defined on objects by the rule

$$C^\infty(\mathbb{R}^n)/I \mapsto C^\infty(\mathbb{R}^n)/\hat{I}$$

with an obvious extension to the morphisms. In particular, $\mathscr{B}$ has finite colimits since $\mathscr{A}_{FT}$ does and they are obtained by means of the reflection.

Remark 9.24 Every point determined C^∞-ring is germ determined. Indeed, if B is point determined, then

$$\widehat{\{0\}} = \{f \in B \mid \forall p \in B \;\; f_p \in \{0\}_p\} = \{f \in B \mid \forall p \in B \;\; f(p) = 0\} = \{0\}.$$

Remark 9.25 The inclusions established in Proposition 9.22 are all proper.

(i) An instance of an object of $\mathscr{B}$ that is not in $\mathscr{A}_{FP}$ is the C^∞-ring $C^\infty_p(M)$ since the germ determined ideal $J(p)$ that defines it is not finitely generated. That it is germ determined is shown as follows. If $f_q \in J(p)_q$ for all $q \in M$ then in particular f_p vanishes in a neighbourhood of p thus $f \in J(p)$. For $M = \mathbb{R}$ and $p \in M$, consider $\{(p-\varepsilon, p+\varepsilon) \mid \varepsilon > 0\}$ and let $\{\xi_\varepsilon\}$ be a set of smooth characteristic functions corresponding to the open subsets $(p-\varepsilon, p+\varepsilon)$. This infinite set of elements of $J(p)$ is not reducible to a finite subset that generates it inside the ideal.

(ii) We point out that not every C^∞-ring of finite type is germ determined. For instance, consider $A = C^\infty(\mathbb{R})/I$ where

$$I = \{f \in C^\infty(\mathbb{R}) \mid \exists \varepsilon > 0 \ \ \forall x \in \mathbb{R}\,(|x| < 0 \Rightarrow f(x) = 0)\}.$$

The ideal I is germ determined so that $A \in \mathscr{B}$. Consider now the standard projection $\pi \colon \mathbb{R}^2 \longrightarrow \mathbb{R}$ onto the x-axis, so that

$$\pi^* I = \{f \in C^\infty(\mathbb{R}^2) \mid \exists \varepsilon > 0\, \forall (x,y) \in \mathbb{R}^2\,(|x| < 0 \rightarrow f(x,y) = 0)\}.$$

We could have $f \in \mathbb{R}^2$ such that for all $(p,q) \in \mathbb{R}^2$, $f_{(p,q)} \in \pi^* I|_{(p,q)}$ yet $f \notin \pi^* I$. Let us be more explicit. The condition that $f_{(p,q)} \in \pi^* I|_{(p,q)}$ says that for some $\varepsilon_p > 0$ there is some g such that

$$\forall (x,y) \in \mathbb{R}^2\,(|x| < \varepsilon_p \rightarrow g(x,y) = 0)$$

and

$$f_{(p,q)} = g_{(p,q)}.$$

Yet, there may not be a single $\varepsilon > 0$ which works for all p, that is, it may happen that $f \notin \pi^* I$.

10
An Application to Unfoldings

As an application of the existence of a well adapted model of SDT we deal with stability of unfoldings of germs as a particular case of that of stability of germs. The power of topos theory is particularly suited for dealing not just with germs but also with their unfoldings. This is an improvement on the classical theory of stability for unfoldings of germs which according to [110] is a separate theory needed to be developed independently from that of stability for germs.

10.1 Wassermann's Theory

It has been claimed by G. Wassermann [110] that the theory of unfoldings of germs at 0 of smooth mappings $\mathbb{R}^n \longrightarrow \mathbb{R}^p$ is necessarily distinct from that of the theory of stable mappings. In particular, the definitions of the main notions involved for unfoldings, such as those of 'r-dimensional equivalence' and of 'stability' are in his view not only seemingly different from their analogues for germs but also quite complicated. Thus, in a separate development, a stability theorem is established in [110] for unfoldings.

We begin by stating the classical notions of Wassermann's theory of unfoldings before introducing their corresponding analogues within SDT.

In [110], an r-dimensional unfolding $\xi \in C_0{}^{\infty}(\mathbb{R}^{n+r})$ is always an unfolding of some germ $\eta \in C_0{}^{\infty}(\mathbb{R}^n, \mathbb{R})$ which need not be mentioned. An r-unfolding ξ is said to be *infinitesimally stable* if for every germ $\omega \in C_0{}^{\infty}(\mathbb{R}^{n+r} \times [0,1], \mathbb{R})$, a germ at $0 \in \mathbb{R}^n \times \mathbb{R}^r \times [0,1]$ of a smooth path beginning at ξ, there exist smooth paths φ, ψ, λ, beginning at $\mathrm{proj}_{\mathbb{R}^n}, \mathrm{proj}_{\mathbb{R}}, \mathrm{id}_{\mathbb{R}^r}$ of germs,

$$\varphi_t \in C_0{}^{\infty}(\mathbb{R}^{n+r}, \mathbb{R}),$$

$$\psi_t \in C_0{}^{\infty}(\mathbb{R}^{1+r}, \mathbb{R}),$$

and

$$\lambda_t \in C_0{}^{\infty}(\mathbb{R}^r, \mathbb{R}^r),$$

for t near 0, such that

$$H(x,u,t) = \psi_t(\omega_t(\varphi_t(x,u), \lambda_t(u)), u)$$

and

$$\frac{\partial H}{\partial t} \equiv 0$$

for $(x,u,t) \in \mathbb{R}^n \times \mathbb{R}^r \times [0,1]$ near $(0,0,0)$.

In the same source [110], an r-dimensional unfolding ξ is called *strongly stable* if for each open neighbourhood $0 \in U \subset \mathbb{R}^{n+r}$ and representative $f \in C^{\infty}(U)$ of ξ, there is an open neighbourhood $f \in W \in C^{\infty}(U)$ (always in the weak topology) such that, for every $g \in W$, there is $(x_0,u_0) \in U \subset \mathbb{R}^n \times \mathbb{R}^r$ and an 'r-dimensional equivalence' (φ, ψ, λ) from the germ of f at (x_0,u_0) to the germ of g at $(\varphi(x_0,u_0), \lambda(u_0))$, which means to give local homeomorphisms φ, ψ, λ at $(x_0,u_0), (g(\varphi(x_0,u_0)), \lambda(u_0)), u_0$ respectively, such that

$$f(x,u) = \psi^{-1}(g(\varphi(x,u), \lambda(u)), u)$$

for all (x,u) in some open neighbourhood of (x_0,u_0).

As an application of Wassermann's stability theorem is the validity of R. Thom's list of the 'seven elementary catastrophes' (in his terminology).

10.2 Unfoldings in SDT

For the purposes of this we shall assume that there exists a well adapted model $(\mathscr{G}, R)$ of SDT with a presentation of the form $\mathscr{G} = \mathrm{Sh}_j(\mathscr{B}^{\mathrm{op}})$ such that the Grothendieck topology j is subcanonical and such that for any $n > 0$, $\Delta(n) = \neg\neg\{0\}$ is representable. Moreover, $f \in C_0{}^{\infty}(\mathbb{R}^{n+r})$ in $\mathscr{G}$ corresponds to the global section $f \in_1 R^{\Delta(n+r)}$.

Definition 10.1 An r-dimensional unfolding of a germ $\eta \in R^{\Delta(n)}$ is a germ $\xi \in R^{\Delta(n+r)}$ such that $\xi|_{\Delta(n)} = \eta$, where $\Delta(n) \longrightarrow \Delta(n+r)$ is given by $x \mapsto (x,0)$, $0 \in \Delta(r)$.

Theorem 10.2 *Let the germ $f \in C_0{}^{\infty}(\mathbb{R}^{n+r})$ in $\mathscr{G}$ correspond to the global section $f \in_1 R^{\Delta(n+r)}$. Then the following two statements hold:*

(i) *f is infinitesimally stable as an r-dimensional unfolding if and only if f is infinitesimally stable as a germ at $0 \in R^n$ of a morphism $R^n \longrightarrow R$ in $\mathscr{G}$.*

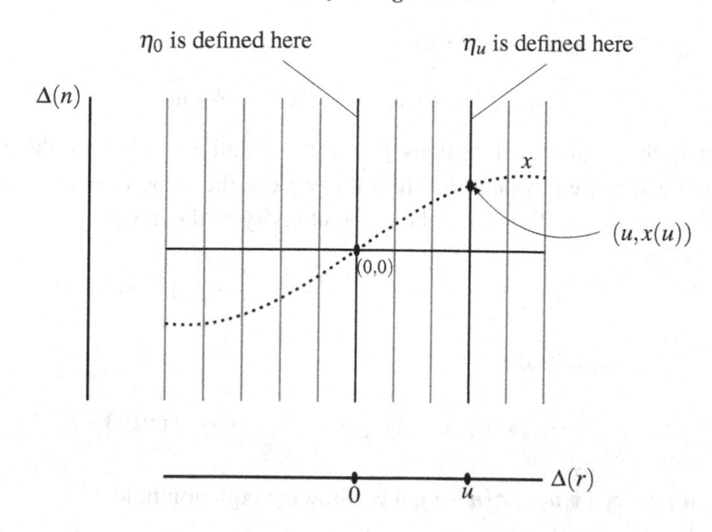

Figure 10.1 Unfolding

(ii) *f is (strongly) stable as an r-dimensional unfolding if and only if f is stable as a germ at $0 \in R^n$ of a morphism $R^n \longrightarrow R$ in $\mathscr{G}$.*

Proof (1). To say that $f \in_{\Delta(r)} R^{\Delta(n)}$ is infinitesimally stable is to say that, given $\omega \in_{\Delta(r)} \text{Vect}(R^n)$ there exists a covering of $\Delta(r)$, necessarily trivial, so reduced to a single isomorphism $\lambda: \Delta(r) \longrightarrow \Delta(r)$, such that there exist $\sigma \in_{\Delta(r)} \text{Vect}(R^n)$ and $\tau \in_{\Delta(r)} \text{Vect}(R)$, such that

$$\models_{\Delta(r)} \lambda^* \omega = \alpha_f(\sigma) \oplus \beta_f(\tau).$$

Given that, in terms of the principal parts of these vector fields regarded as $\omega: \Delta(n+r) \longrightarrow R$ with $\sigma|_0: \Delta(n+r) \longrightarrow R^n$ and $\tau|_0: \Delta(1+r) \longrightarrow R$, the above translates to infinitesimal stability stated for unfoldings by an argument that is standard and has already been used in Step 4 of the first proof of Mather's theorem (Theorem 7.11), the claim is true. Notice also, for this, that $\lambda^* \omega(x,u) = \omega(x, \lambda(u))$.

(2). To say that $f \in_{\Delta(r)} R^{\Delta(n)}$ is stable is to say that $\text{Im}\,(\gamma_f) \in_{\Delta(r)} P(R^{\Delta(n)})$. Now,

$$\text{Im}\,\gamma_f = [[g \in_{\Delta(r)} R^{\Delta(n)} \mid \exists \varphi \in_{\Delta(r)} \text{Inf.inv}_0(R^{nR}) \exists \psi \in_{\Delta(r)} \text{Inf.inv}_0(R^R) \\ (g = \psi|_0 \circ f \circ (\varphi|_0)^{-1} \circ \alpha_{\varphi(0)})]].$$

Said in other words, $g \in_{\Delta(r)} \text{Im}\,\gamma_f$ if there exists a covering of $\Delta(r)$, that is to say, an isomorphism $\lambda: \Delta(r) \longrightarrow \Delta(r)$, together with $\varphi \in_{\Delta(r)} \text{Inf.inv}_0(R^{nR^n})$

and $\psi \in_{\Delta(r)} \text{Inf.inv}_0(R^R)$ such that

$$\models_{\Delta(r)} \lambda^* g = \psi|_0 \circ f \circ (\varphi|_0)^{-1} \circ \alpha_{\varphi(0)}.$$

Now, in terms of global sections $f \in_1 R^{\Delta(n+r)}$ and $g \in_1 R^{\Delta(n+r)}$, the condition can be stated equivalently as follows (we use the same notations, $\varphi|_0 \in_1 R^{n\Delta(n+r)}$ and $\psi|_0 \in_1 R^{\Delta(1+r)}$) as the commutativity of the diagram

$$\begin{array}{ccc} \Delta(n+r) & \xrightarrow{\langle f, \pi_r^{n+r} \rangle} & \Delta(1+r) \\ \Big\downarrow{\scriptstyle \langle \varphi|_0, \lambda|_0 \circ \pi_r^{n+r} \rangle} & & \Big\downarrow{\scriptstyle \psi|_0} \\ \neg\neg\{(\varphi(0), \lambda(0))\} & \xrightarrow[g \circ \alpha_{\langle \varphi|_0, \lambda|_0 \circ \pi_r^{n+r} \rangle}]{} & \neg\neg\{\psi(0)\} \end{array}$$

that is, for every $(x,u) \in \Delta(n+r)$, the following equation holds:

$$f(x,u) = \overline{\psi}^{-1}(g(\varphi(x,u), \lambda(u)), u)$$

where $\overline{\psi} \colon \Delta(1+r) \longrightarrow \neg\neg\{\psi(o)\} \times \Delta(r)$ is given by $\overline{\psi} =< \psi|_0, \pi_r^{1+r} >$, which clearly shows the equivalence with the classical notion. □

It is now possible to obtain Theorem 4.11 of [110] as a corollary of Mather's theorem if the latter is interpreted in $\mathscr{G}$. Indeed, Wassermann's theorem on the equivalence of stability and infinitesimal stability of unfoldings is just a consequence of the validity of Mather's theorem in any model of SDT and of the existence of a well adapted model of SDT. We state it below.

Theorem 10.3 (Wassermann's theorem) *[110] Let ξ be an r-dimensional unfolding of a germ η of a smooth mapping $\mathbb{R}^n \longrightarrow \mathbb{R}$. Then ξ is (strongly) stable if and only if ξ is infinitesimally stable.*

PART SIX

A WELL ADAPTED MODEL OF SDT

In this sixth part we deal with the construction of a well addapted model of SDT. There are several known models of SDG, some that are well adapted (in a technical sense [10]) for applications to classical differential geometry and analysis. A survey of the models of SDG is given in [61] and [90]. We shall single out from the start just one of these models—to wit, the Dubuc topos $\mathscr{G}$ [34] together with an object R in it, as it is the only one known to us to be also a model of the axioms for SDT as presented in this book. The model $(\mathscr{G}, R)$ is one of several built from C^∞-rings. Although Grothendieck toposes in general (and the topos $\mathscr{G}$ in particular) provide natural models of SDG, hence are suitable for the intended applications to classical mathematics, there are good reasons for developing SDG within an arbitrary topos. Indeed, by so doing, all constructions done in that setting are free from any given set theory. Moreover, working within a topos is tantamount to doing so constructively.

11
The Dubuc Topos $\mathscr{G}$

In this chapter we recall the construction of the Dubuc topos $\mathscr{G}$ by means of C^∞-rings and germ determined ideals.

11.1 Germ Determined Ideals of C^∞-Rings

In this section we use notions that were introduced in § 9.1, in particular those of a C^∞-ring and of a germ determined ideal of it. Recall that $\mathscr{B}$ denotes the full subcategory of the category $\mathscr{A}$ of C^∞-rings whose objects are those C^∞-rings of finite type of the form $B = C^\infty(\mathbb{R}^n)/J$ where $J \subset C^\infty(\mathbb{R}^n)$ is a germ determined ideal. Recall in particular that several objects of $\mathscr{G}$ that will be relevant in what follows are representable. For instance, this is the case for the objects R, of any intrinsic open V of some R^m, and of the closed interval $[0,1]$. We will establish below that also the object $\Delta = \neg\neg\{0\}$ is representable, as are all the $\Delta(n) \subset R^n$.

Definition 11.1 The (Dubuc) topos $\mathscr{G}$ [34] is defined as the topos of sheaves on $\mathscr{B}^{\mathrm{op}}$ with the *open cover (Grothendieck) topology* on it—the latter described, dually, on $\mathscr{B}$, as that generated by countable families

$$\{C^\infty(\mathbb{R}^n) \longrightarrow C^\infty(U_\alpha)\}$$

where $\{U_\alpha\}$ is an open covering of $\mathbb{R}^n$, plus the empty family cocovering 0.

Regardless of the nature of the ideals of definition of the C^∞-rings in $\mathscr{A}_{FT}$, the open cover topology is meaningful and its cocovers are families of the form

$\{A \longrightarrow A\{a_\alpha{}^{-1}\}\}$ obtained by pushout

$$\begin{array}{ccc} C^\infty(\mathbb{R}^s) & \longrightarrow & C^\infty(U_\alpha) \\ {\scriptstyle\varphi}\downarrow & & \downarrow \\ A & \longrightarrow & A\{a_\alpha{}^{-1}\} \end{array}$$

where $a_\alpha = \varphi(g_\alpha)$ and g_α a smooth characteristic function of U_α with $\{U_\alpha\}$ covering $\mathbb{R}^s$.

Recall that a Grothendieck topology is *subcanonical* if every representable functor is a sheaf. Equivalently, this is the case if the covers are (pullback) stable effective epimorphic families. In dual terms, this says that the cocovers should be (pushout) stable effective monomorphic families. Explicitly, for a cocover

$$\{A \longrightarrow A_\alpha\},$$

where $A_\alpha = A\{a_\alpha^{-1}\}$, it should be the case that given any compatible family $\{b^\alpha \in A\{a_\alpha^{-1}\}\}$ there be a unique $b \in A$ such that $b_\alpha = b^\alpha$.

Remark 11.2 Here is an example that shows that in $\mathscr{A}_{FP}{}^{\mathrm{op}}$, the open cover topology is not subcanonical. Consider the cocover $\{A \longrightarrow A\{a_\alpha^{-1}\}\}$ with $A = C^\infty(\mathbb{R})/I$, where $I = \{\varphi \mid \mathrm{supp}(\varphi)\,\mathrm{is\,compact}\}$ and let $\{U_\alpha\}$ be an open cover corresponding to a_α, that is, such that $U_\alpha = \varphi^{-1}(\mathbb{R}^*)$. It is possible to assume that for each α, $\varphi_\alpha \in I$, that is, that φ_α has compact support. It follows that $a_\alpha = [\varphi_\alpha] = 0$ so that $A\{a_\alpha^{-1}\} = 0$ for each α. Since $A \neq 0$, $A \longrightarrow 0$ is not monomorphic. Of course the original cocoverings of the form $\{C^\infty(\mathbb{R}^n) \longrightarrow C^\infty(U_\alpha)\}$ where $\{U_\alpha\}$ is an open covering of $\mathbb{R}^n$ are effective monomorphic (given a compatible family of smooth functions g_α defined on the U_α lift to $C^\infty(\mathbb{R}^n)$ uniquely) but are not universal, as the above example shows.

Proposition 11.3 *The open cover Grothendieck topology on $\mathscr{B}^{\mathrm{op}}$ is subcanonical.*

Proof We need to show that the cocovers $\{A \longrightarrow A_\alpha = A\{a_\alpha^{-1}\}\}$ are effective monomorphic, or that for each such the diagram

$$A \longrightarrow \prod_\alpha A_\alpha \rightrightarrows \prod_{\alpha,\beta} A_{\alpha\beta}$$

is an equalizer, where

$$A_{\alpha\beta} = A\{a_\alpha^{-1}, a_\beta^{-1}\} = A\{a_\alpha^{-1}\}\{a_\beta^{-1}\}.$$

Let $\{b^\alpha \in A_\alpha\}$ be a compatible family. Then, for each point p of $A_{\alpha\beta}$, we have $(b^\alpha)_p = (b^\beta)_p$. Define $b(p) \in A_p$ by means of

$$(b(p))_\alpha = (b^\alpha)_p$$

for any α such that p is a point of A_α. (Note: one such must be the case by the covering property and well defined by the compatibility.)

Claim: There exists $b \in A$ with

$$b_p = b(p)$$

for each point p of A.

Before proving the claim notice that if true it would imply that $b_\alpha = b^\alpha$, where we recall the notation b_α indicates the image of b in $A\{a_\alpha^{-1}\}$, that is, the germ of b at α. Since $A\{a_\alpha^{-1}\}$ is germ determined it is enough that for each point p of $A\{a_\alpha^{-1}\}$, $(b_\alpha)_p = (b^\alpha)_p$. This is the case as

$$(b_\alpha)_p = (b_p)_\alpha = (b(p))_\alpha = (b^\alpha)_p$$

where the first identity is always true by a commutative diagram, the second identity is the claim, and the third identity is by definition.

We now prove the claim. Let $h_\alpha \in C^\infty(U_\alpha)$, $b_\alpha = [h_\alpha]$, and $\{W_i\}$ a locally finite refinement of the $\{U_\alpha\}$, for i let α such that $W_i \subset U_\alpha$, $g_i \in C^\infty(W_i)$ and $g_i = h_\alpha|_{W_i}$. Remark that for all $p \in W_i$,

$$[(g_i)_p] = b(p)$$

since

$$b(p) = (b_\alpha)_p = [(h_\alpha)_p] = [(g_i)_p].$$

Now let $\{\varphi_i\}$ be a subordinated partition of unity. The functions

$$l_i = \varphi_i \cdot g_i$$

are globally defined since $g_i \in C^\infty(W_i)$ and $\mathrm{supp}(\varphi)_i \subset W_i$, $\mathrm{supp}(l_i) \subset W_i$. Thus, given any point p of A, if $p \notin W_i$ then $(l_i)_p = 0$ and if $p \in W_i$ then

$$[(l_i)_p] = [(\varphi_i)_p] \cdot b(p).$$

Since the family $\{l_i\}$ is locally finite, $l = \sum l_i$ exists. Let

$$b = [l].$$

Now, for any point p of A, and for the $p \subset W_i$,

$$b_p = [\sum_i (h_i)_p] = \sum_i [(l_i)_p] = b(p) \cdot \sum (\varphi_i)_p = b(p)$$

since $\sum_i (\varphi_i)_p = 1$. □

By construction, $\mathscr{G} = \mathrm{Sh}(\mathscr{B}^{\mathrm{op}}) \hookrightarrow \mathit{Set}^{\mathscr{B}}$. By Proposition 11.3, the Yoneda embedding factors through this inclusion:

$$\mathscr{B}^{\mathrm{op}} \xrightarrow{\text{yon}} \mathscr{G} \hookrightarrow \mathit{Set}^{\mathscr{B}}$$

and this functor preserves finite limits and open coverings, therefore the restriction to $\mathscr{M}^{\infty}$, that is the composite $\iota : \mathscr{M}^{\infty} \hookrightarrow \mathscr{G}$ given by

$$\mathscr{M}^{\infty} \xrightarrow{C^{\infty}(-)} \mathscr{B}^{\mathrm{op}} \xrightarrow{\text{yon}} \mathscr{G}$$

preserves transversal pullbacks, 1, and open coverings. This observation constitutes a proof of the following.

Theorem 11.4 *The pair $(\mathscr{G}, R)$, with $\mathscr{G} = \mathrm{Sh}(\mathscr{B}^{\mathrm{op}})$ and $R = \iota(\mathbb{R})$, where $\iota : \mathscr{M}^{\infty} \hookrightarrow \mathscr{G}$ is as defined above, is a well adapted model of ringed toposes.*

As already recalled above, among the representable functors (which are sheaves) are the objects indicated below, where the bar indicates the corresponding representable functor:

$$R = \overline{C^{\infty}(\mathbb{R})},$$
$$D = \overline{C^{\infty}(\mathbb{R})/(t^2)}\,.$$

This is also the case in other models of SDG. In $\mathscr{G}$ we have moreover the following important result:

Proposition 11.5 *In the topos $\mathscr{G}$, the object $\Delta(n) = \neg\neg\{0\} \subset R^n$ is representable by the ring $C_0^{\infty}(\mathbb{R}^n)$.*

Proof By definition, $\overline{C_0^{\infty}(\mathbb{R}^n)}$ is the external intersection of the $\overline{C^{\infty}(U)}$ for all open sets $0 \in U \subset \mathbb{R}^n$. From Proposition 11.12 follows that $P_0(R^n) \subset \overline{C_0^{\infty}(\mathbb{R}^n)}$. The converse follows from Corollary 5.20 [25, Lemma 1.7]. Let $U \in_{\overline{A}} P(R^n)$ where $A = C_0^{\infty}(\mathbb{R}^n)$. For the corresponding subobject $U \subset R^n \times \overline{A}$ we have that $\{0\} \times \overline{A} \subset U$. By Corollary 5.20 there exists an intrinsic open H of $R^n \times \overline{A}$ such that $\{0\} \times \overline{A} \subset H \subset U$. In particular, $\{0\} \times Z(I) \subset \Gamma(H)$. But $\{0\} \times Z(I) = Z(J) \times Z(I) = Z(J, I)$ where J denotes the ideal of functions in n variables of null germ at the origin.

Thus, by Proposition 11.12, it follows that

$$C_0^{\infty}(\mathbb{R}^n) \times \overline{A} \subset H.$$

This finishes the proof. □

Remark 11.6 It is easy to see that the object $C^{\infty}{}_0(\mathbb{R}^n)$ of $\mathscr{B}$ has only trivial cocoverings. Therefore, $\Delta(n) = \overline{C^{\infty}{}_0(\mathbb{R}^n)}$ has only trivial covers in $\mathscr{B}^{\mathrm{op}}$.

In addition, the usual strict order on $\mathbb{R}$ induces one on R defined by $R_{>0} = \iota(\mathbb{R}_{>0})$. To defined closed intervals several choices are presented since they are not objects of $\mathscr{M}^\infty$. Define

$$R_{\geq 0} = \overline{C^\infty(\mathbb{R})/\mathfrak{m}^o_{[0,\infty)}}$$

where $\mathfrak{m}^o_X$ is the ideal of functions vanishing on X. This gives a preorder compatible with the ring structure, total on the invertibles, and for which the closed interval $[0,0]$ of R contains the nilpotents.

Remark 11.7 The condition $x \geq y$ implies $\neg(x < y)$ but not conversely. In fact,

$$\neg R_{<0} = \overline{C^\infty(\mathbb{R})/m^g_{[0,\infty)}}$$

where m^g_X is the ideal of functions vanishing on *an open neighbourhood* of X.

The following is a useful result about $\mathscr{G}$.

Proposition 11.8 *(Nullstellensatz) Consider the global sections functor*

$$\Gamma\colon \mathscr{G} \longrightarrow \mathit{Set}.$$

Then, for any $X \in \mathscr{G}$, $\Gamma(X) = \emptyset$ if and only if $X = 0$.

Proof Assume that X is such that $\Gamma(X) = \emptyset$. Let $A_\alpha = C^\infty(\mathbb{R}^n)/I_\alpha \in \mathscr{B}$. Denote by $\overline{A_\alpha}$ the corresponding representable functor, which is a sheaf on $\mathscr{B}^{\mathrm{op}}$ for the open cover topology. Since the representables generate, there must be some epimorphic family $\{\overline{A_\alpha} \longrightarrow X\}$ and therefore one for which $\Gamma(\overline{A_\alpha}) = \emptyset$ for all α. Since I is germ determined it follows trivially from this that $I_\alpha = C^\infty(\mathbb{R}^n)$ and therefore $A_\alpha = 0$. Therefore, $X = 0$. The converse implication is trivially true. □

We will also need the following result in the next section.

Proposition 11.9 *Let W be a Weil algebra over $\mathbb{R}$, $B \in \mathscr{B}$. Then $B \otimes_\infty W \in \mathscr{B}$.*

Proof As $\pi\colon W \longrightarrow \mathbb{R}$ is the unique point of W, there is a bijection between points of $B \otimes_\infty W$ and points of B via the coproduct inclusion $B \hookrightarrow B \otimes_\infty W$. Now,

$$\begin{aligned} B \otimes_\infty W \simeq B \otimes_{\mathbb{R}} W &= B \otimes_{\mathbb{R}} (\mathbb{R} \oplus W') \\ &\simeq (B \otimes_{\mathbb{R}} \mathbb{R}) \oplus (B \otimes_{\mathbb{R}} W') \simeq B \oplus (B \otimes_{\mathbb{R}} W') \end{aligned}$$

where all elements in the second summand are nilpotent.

In particular if $b = b_0 + n \in B \oplus (B \otimes_{\mathbb{R}} W')$, then

$$(B \otimes_\infty W)\{b^{-1}\} = B\{b_0{}^{-1}\} \otimes_\infty W$$

since n is nilpotent. It follows that for any point $\bar{p}$ of B with $p = \bar{p} \cdot \bar{\pi}$ the corresponding point of $B \otimes_\infty W$, we have

$$(B \otimes_\infty W)_p \simeq B_{\bar{p}} \otimes_\infty W.$$

To say that B is germ determined is to say that the family $\{B \longrightarrow B_{\bar{p}}\}$ for all points $\bar{p}$ of B, is jointly monic. Now, the functor $(-) \otimes_{\mathbb{R}} W = (-) \otimes_\infty W$ preserves jointly monic families since, on the underlying vector spaces, it is of the form $C \mapsto C \oplus \cdots \oplus C$ (m times, where m is the linear dimension of W over $\mathbb{R}$, that is, if $W \simeq \mathbb{R}^m$). It follows that

$$\{B \otimes_\infty W \longrightarrow B_{\bar{p}} \otimes_\infty W \simeq (B \otimes_\infty W)_p\}$$

is jointly monic for each point p of $B \otimes_\infty W$, hence the latter is germ determined. □

Remark 11.10 On account of the reflection to the inclusion

$$\mathscr{B} \hookrightarrow \mathscr{A}_{FT},$$

the category $\mathscr{B}$ has all limits and colimits that exist in $\mathscr{A}_{FT}$. However, and as already noted, coproducts in $\mathscr{B}$ are in general obtained by reflection of coproducts in $\mathscr{A}_{FT}$. The case considered in Proposition 11.9 is special.

Similarly, localizations in $\mathscr{B}$ need not be the same as those in $\mathscr{A}$. For instance, if $A = C^\infty(\mathbb{R}^n)/I$ with I germ determined, and $a \in A$,

$$A\{a^{-1}\} = C^\infty(\mathbb{R}^{n+1}/(\pi^* I, 1 - \varphi(\bar{x}) \cdot y))$$

which may not be in $\mathscr{B}$ as $\pi^* I$ need not be germ determined. An example is given by

$$I = \{f \in C^\infty(\mathbb{R}) \mid \exists \varepsilon > 0 \ \forall |x| < \varepsilon \ f(x) = 0\}$$

since

$$\pi^* I = \{f \in C^\infty(\mathbb{R}^2) \mid \exists \varepsilon > 0 \ \forall |x| < \varepsilon \ \forall y \in \mathbb{R} \ f(x,y) = 0\}$$

is not germ determined. The argument to show this is the same as that employed in Remark 9.25.

We end this chapter with some remarks concerning the connection between the intrinsic topological structure for certain objects of $\mathscr{G}$ and the topological structures used in classical analysis. Several such results have been obtained by J. Penon [96] and E. Dubuc [35]. Among them we quote the following.

Proposition 11.11 *Let $\overline{A}$ be any representable in $\mathscr{G}$, say $A = C^\infty(\mathbb{R}^n)/I$. Let $X \subset \overline{A}$. Then, $X \in P(\overline{A})$ if and only if $\Gamma(X) \subset \Gamma(\overline{A}) = Z(I)$ is open in the usual sense. More precisely, there exists a right adjoint $\Gamma \vdash \Lambda$ such that $\Gamma\Lambda =$ id.*

For particular kinds of objects, the bijection implicit in Proposition 11.11 admits a concrete interpretation. We quote:

Proposition 11.12 *[96, 35] For any object of the form ιM, Γ and ι establish a bijection between intrinsic open parts of ιM and (classical) open subsets of M.*

We now give a characterization of intrinsic opens of $R^{\overline{A}}$ in terms of the global sections functor $\Gamma\colon \mathscr{G} \longrightarrow$ *Set*. Unless otherwise stated, these results are from [44] (also included in [26]).

Lemma 11.13 *Let X be an object of $\mathscr{G}$, $i\colon U \rightarrowtail X$. Then $U \in P(X)$ if and only if for every representable object $\overline{B}$ —with B say of the form $C^\infty(\mathbb{R}^m)/K$— and morphism $\alpha\colon \overline{B} \longrightarrow X$, in the pullback*

$$\begin{array}{ccc} W & \longrightarrow & \Gamma(U) \\ \downarrow & \lrcorner & \downarrow{\scriptstyle \Gamma(i)} \\ \Gamma(\overline{B}) & \xrightarrow[\Gamma(\alpha)]{} & \Gamma(X) \end{array}$$

we have that $W \in P(\overline{B}) = E(\overline{B})$, that is, W is open for the Euclidean topology of $\Gamma(\overline{B}) \subset \mathbb{R}^m$.

Proof The condition that is claimed $U \in P(X)$ may be equivalently stated as follows: for every $\overline{B} = C^\infty(\mathbb{R}^m)/K$ in $\mathscr{B}$, and $\alpha\colon \overline{B} \longrightarrow X$, if $b\colon \mathbf{1} \longrightarrow \overline{B}$ is such that $\alpha \circ b \in U$, then there is a neighbourhood V of $b \in R^m$ such that $b \in \overline{V}$ and $\alpha[\overline{V}] \subset U$.

On the other hand, $i\colon U \rightarrowtail X$ is intrinsic open means that the statement

$$\forall f \in U \forall g \in X [\neg(g = f) \vee g \in U]$$

so that, for instance, for any $\alpha\colon \overline{B} \longrightarrow X, b\colon \mathbf{1} \longrightarrow \overline{B}$, one has that $\alpha \circ b \in U$. In particular,

$$\forall g \in X [\neg(g = \alpha \circ b) \vee g \in U],$$

that is,

$$X = \neg\{\alpha \circ b\} \cup U.$$

Now, $\alpha \in_{\overline{B}} X$, therefore

$$\models_{\overline{B}} \neg(\alpha = \alpha \circ b) \vee \alpha \in U$$

and hence there is a covering family $\{\overline{B_i} \longrightarrow \overline{B}\}_{i \in I}$ such that for each $i \in I$ either $\models_{\overline{B}} \neg(\alpha = \alpha \circ b)$ or $\models_{\overline{B}} \alpha \in U$.

Applying global sections, we obtain a surjective family

$$\{\Gamma(\overline{B_i}) \longrightarrow \Gamma(\overline{B})\}$$

with $\{V_i\}$ an open covering of $\mathbb{R}^m$, such that $\{V_i \cap Z(K)\}$ is an open covering of $Z(K)$, the set of zeroes of K in $\mathbb{R}^m$. Since $b \in \Gamma(\overline{B})$, it must be the case that for some $i_0 \in I$, $b \in \Gamma(\overline{B_{i_0}})$. We claim that $\neg \models_{\overline{B_{i_0}}} \neg(\alpha = \alpha \circ b)$ since, for $b \colon \mathbf{1} \longrightarrow \overline{B_{i_0}}$, $\models_{\mathbf{1}} \alpha \circ b = \alpha \circ b$. Therefore,

$$\models_{\overline{B_{i_0}}} \alpha \in U.$$

Let $V = V_{i_0}$ so that

$$\overline{B_{i_0}} = \overline{C^\infty(V_{i_0})/(K/V_{i_0})} = \overline{V_{i_0}}$$

is as required. □

On the basis of Lemma 11.13, I. Moerdijk found a proof of the following fact relating intrinsic opens of $R^{\overline{A}}$ in $\mathscr{G}$ and weak C^∞-opens of $\Gamma(R^{\overline{A}})$ (private communication; see also [14] where the theorem was found independently and completed to include the reverse implication).

Proposition 11.14 *Let $\overline{A}$ be representable in $\mathscr{G}$, say $A = C^\infty(\mathbb{R}^n)/I$ in $\mathscr{B}$. Then, for $i \colon U \rightarrowtail R^{\overline{A}}$, if $U \in P(R^{\overline{A}})$ then $\Gamma(U) \subset C^\infty(\mathbb{R}^n)/I$ is open in the (quotient) weak topology.*

Proof We deal with the case $\overline{A} = \mathbb{R}^n$ as the general case is similar. The proof is based on the following characterization of opens in the weak C^∞-topology. A subset $V \subset C^\infty(\mathbb{R}^n)$ is open in the weak C^∞-topology if for every smooth path

$$[0,1] \xrightarrow{F} C^\infty(\mathbb{R}^n)$$

$F^{-1}(V)$ is open in $[0,1]$ for the induced Euclidean topology. Since smooth operators between Frechet spaces are continuous for the Frechet topology [43], the condition is necessary.

Conversely, assume $V \subset C^\infty(\mathbb{R}^n)$ not open. In that case there must exist some sequence $\{f_m\} \longrightarrow f$, convergent in the weak C^∞-topology, with $f \in V$ and $f_m \notin V$ for all $m > 0$. By a result in [99], there is a subsequence $\{f_{m_k}\} \longrightarrow f$, and a smooth mapping $F \colon [0,1] \longrightarrow C^\infty(\mathbb{R}^n)$ with $F(0) = f$ and $F(\frac{1}{k}) = f_{m_k}$ for each $k > 0$. By assumption we have $0 \in F^{-1}(V)$ but $\frac{1}{k} \notin F^{-1}(V)$ for all $k > 0$. This shows that $F^{-1}(V)$ is not open in $[0,1]$.

To conclude, notice that, in $\mathscr{G}$, $[0,1] = \overline{C^\infty(\mathbb{R})/\mathrm{m}^g{}_{[0,1]}}$ and that a smooth map $F \colon [0,1] \longrightarrow C^\infty(\mathbb{R}^n)$ as above lifts to a morphism $\alpha \colon [0,1] \longrightarrow R^{R^n}$ in $\mathscr{G}$. (In order to handle the case $R^{\overline{A}}$, reduce it to the above by using that in the quotient topology of $C^\infty(\mathbb{R}^n)/I$, $\{[f_n]\} \longrightarrow [f]$ if and only if there exists a sequence $\{g_n\} \longrightarrow g$ in $C^\infty(\mathbb{R}^n)$ with $[f_n] = [g_n]$ and $[f] = [g]$.) Now, use Lemma 11.13 to finish the proof. □

The reverse implication is dealt with by O. Bruno [14]. In the following we extract from it what is needed to complete the above.

Proposition 11.15 *(a) Let X be any object of $\mathscr{G}$. Then, the correspondence*

$$U \subset X \mapsto \Gamma(U) \subset \Gamma(X)$$

from subobjects of X to subsets of $\Gamma(X)$ (evidently functorial) has a right adjoint Λ described as follows: for $S \subset \Gamma(X)$, let $\Lambda(S) \subset X$ be given by the rule: for any representable $\overline{B}$ and $\alpha\colon \overline{B} \longrightarrow X$, α factors through $\Lambda(S) \subset X$ if and only if $\Gamma(\alpha)$ factors through $S \subset \Gamma(X)$. It is always the case that $\Gamma\Lambda(S) = S$.

(b) If $U \subset X$ is intrinsic open, then $\Lambda(\Gamma(U)) = U$.

(c) Let $V \subset \Gamma(R^{\overline{A}})$ be open in the weak C^∞-topology. Then $\Lambda(V) \subset R^{\overline{A}}$ is intrinsic open.

Proof (a) The definition of $\Lambda(S)$ gives a presheaf on $\mathscr{B}^{\mathrm{op}}$ — it can be checked that it is a sheaf for the open cover topology.

(b) It is enough to show that, for any representable $\overline{B}$, a functor $\alpha\colon \overline{B} \longrightarrow X$ factors through $U \subset X$ if and only if it factors through $\Gamma(U) \subset \Gamma(X)$. Since $U \subset X$ is intrinsic open, it follows from Lemma 11.13 that there exists an open $V \subset \mathbb{R}^m$ (where $B = C^\infty(\mathbb{R}^m)/K)$) such that $\alpha|_{\overline{V}}$ factors through $U \subset X$. It follows that there exists an open covering $\{\overline{V_i}\}$ of $\overline{B}$, on each portion of which α factors through $U \subset X$. Since U is a sheaf, α factors through $U \subset X$.

(c) Here we check the condition given in Lemma 11.13 under the hypothesis. Let $b\colon \mathbf{1} \longrightarrow \overline{B}$ be such that $\alpha \circ b \in \Lambda(V) \subset R^{\overline{A}}$. By definition of $\Lambda(V)$ it follows that $\alpha \circ b \in V$. Now, if $A = C^\infty((R)^n)/I$, α is represented by a smooth map $F\colon \mathbb{R}^{m+n} \longrightarrow \mathbb{R}$ defined modulo $\widetilde{(K,I)}$ and so, $\alpha \circ b$ is represented by $F(b,-)\colon \mathbb{R}^n \longrightarrow \mathbb{R}$ defined modulo I.

Since V is open and the class $[F(b,-)] \in V$, this is so for every b in a neighbourhood $U \cap Z(K)$ of b in $Z(K)$, $U \subset \mathbb{R}^m$ open. We have that $\alpha \circ j_U$ factors through $\Lambda(V) \subset R^{\overline{A}}$.

□

11.2 The Topos $\mathscr{G}$ as a Model of SDG

Let

$$\mathscr{G} = \widetilde{\mathscr{B}^{\mathrm{op}}} = \mathrm{Sh}(\mathscr{B}^{\mathrm{op}})$$

be the Dubuc topos , and let

$$\imath \colon \mathscr{M}^\infty \hookrightarrow \mathscr{G}$$

be the composite

$$\mathscr{M}^\infty \xrightarrow{C^\infty(-)} \mathscr{B}^{\mathrm{op}} \xrightarrow{\text{yon}} \mathscr{G}.$$

Let W be a Weil algebra, say of the form $\mathscr{C}^\infty(\mathbb{R}^n)/(f_1,\ldots,f_k)$. We have $\imath W = \overline{C^\infty(W)}$ the representable functor corresponding to $W \in \mathscr{A}_{FP} \subset \mathscr{B}$ and so it is actually in $\mathscr{G}$ as it is a sheaf (subcanonical topology).

Define

$$jW = \mathrm{Spec}_R(W) = [[\bar{a} \in R^n \mid \bigwedge f_i(\bar{a}) = 0]]$$

where $W = \mathbb{R}[X_1,\ldots,X_n]/(f_1,\ldots,f_k)$ is a presentation.

Lemma 11.16

$$jW \simeq \imath W.$$

Proof In order to prove it we use the generating property in $\mathscr{G}$ of the objects $B \in \mathscr{B}$ regarded as representable functors $\overline{B}$, since those are sheaves by Proposition 11.3. There are given bijections, natural in B:

$$\{\overline{B} \longrightarrow jW = [[\bar{a} \in R^n \mid \bigwedge f_i(\bar{a}) = 0]]\}$$

$$\overline{\qquad\qquad\qquad\qquad\qquad\qquad}$$

$$\{\overline{B} \xrightarrow{\bar{a}} R^n \mid \bigwedge f_i(\bar{a}) = 0\}$$

$$\overline{\qquad\qquad\qquad\qquad\qquad\qquad}$$

$$\{C^\infty(\mathbb{R}^n) \xrightarrow{\varphi} B \mid \bigwedge \varphi(f_i) = 0\}$$

$$\overline{\qquad\qquad\qquad\qquad\qquad\qquad}$$

$$\{C^\infty(\mathbb{R}^n)/(f_1,\ldots,f_k) \xrightarrow{\varphi} B\}$$

$$\overline{\qquad\qquad\qquad\qquad\qquad\qquad}$$

$$W \xrightarrow{\varphi} B$$

$$\overline{\qquad\qquad\qquad\qquad\qquad\qquad}$$

$$\overline{B} \longrightarrow \overline{W}$$

where the first two lines are morphisms in $\mathscr{G}$ and the remaining lines are morphisms in $\mathscr{B}$. □

Theorem 11.17 *$(\mathscr{G}, R)$ satisfies* **Axiom J.**

Proof In view of Proposition 11.3 and Lemma 11.16, Axiom J holds for the pair $(\mathscr{G}, R)$ if and only if, for any Weil algebra W, the canonical morphism

$$iW \xrightarrow{\alpha} R^{\overline{W}}$$

in $\mathscr{G}$ is an isomorphism. We prove it by indicating a sequence of bijections between morphisms with domain an arbitrary representable $\overline{B}$ for $B \in \mathscr{B}$, natural in B. The details rely on previously established statements and are left to the reader.

$$\overline{B} \longrightarrow R^{\overline{W}}$$
$$\overline{\qquad\qquad\qquad\qquad\qquad\qquad}$$
$$\overline{B} \times \overline{W} \longrightarrow R$$
$$\overline{\qquad\qquad\qquad\qquad\qquad\qquad}$$
$$\overline{B \otimes_\infty W} \longrightarrow \overline{C^\infty(\mathbb{R})}$$
$$\overline{\qquad\qquad\qquad\qquad\qquad\qquad}$$
$$C^\infty(\mathbb{R}) \longrightarrow B \otimes_\infty W \simeq B^m$$
$$\overline{\qquad\qquad\qquad\qquad\qquad\qquad}$$
$$\overline{b} \in B^m$$
$$\overline{\qquad\qquad\qquad\qquad\qquad\qquad}$$
$$C^\infty(\mathbb{R}) \xrightarrow{b_1} B, \ldots, C^\infty(\mathbb{R}) \xrightarrow{b_m} B$$
$$\overline{\qquad\qquad\qquad\qquad\qquad\qquad}$$
$$C^\infty(\mathbb{R}) \otimes_\infty \cdots \otimes_\infty C^\infty(\mathbb{R}) \longrightarrow B$$
$$\overline{\qquad\qquad\qquad\qquad\qquad\qquad}$$
$$C^\infty(\mathbb{R}^m) \longrightarrow B$$
$$\overline{\qquad\qquad\qquad\qquad\qquad\qquad}$$
$$C^\infty(W) \longrightarrow B$$
$$\overline{\qquad\qquad\qquad\qquad\qquad\qquad}$$
$$\overline{B} \longrightarrow iW$$

□

Theorem 11.18 *$(\mathscr{G}, R)$ satisfies* **Axiom W.**

Proof Axiom W holds in $\mathscr{G}$ if and only if for every Weil algebra W, the object jW of $\mathscr{G}$ is tiny. This means that the endofunctor

$$(-)^{jW} : \mathscr{G} \longrightarrow \mathscr{G}$$

has a right adjoint and that jW is well supported. Recall that by Lemma 11.16,

$jW \simeq \iota W$. Since W is a Weil algebra it has a (unique) point $\pi\colon W \longrightarrow \mathbb{R}$, hence it is well supported. Every representable $\overline{W}$ is an atom in $Set^{\mathscr{B}}$ because $\mathscr{B}$ has coproducts with Weil algebras and the topos of presheaves on $\mathscr{B}^{\text{op}}$ is cocomplete and has a small set of generators. It follows from Theorem 1.32 and the fact that the open cover topology is subcanonical, that $\overline{W}$ is an atom in $\mathscr{G} = \text{Sh}(\mathscr{B}^{\text{op}})$. Therefore $\overline{W}$ is a tiny object in $\mathscr{G}$. □

Remark 11.19 The proof above of the validity of Axiom W in $\mathscr{G}$ relies on some general results from topos theory. It can also be verified directly as follows, where we use not just that the representable functor corresponding to a Weil algebra is well supported but that in fact it has a unique point.

Recall that the associated sheaf functor

$$a\colon Set^{\mathscr{B}} \longrightarrow \text{Sh}(\mathscr{B}^{\text{op}})$$

is the composite $a = l \circ l$, where

$$l(X)(B) = \text{colim}_{\mathscr{R}(B)} \text{Hom}(\mathscr{R}(B), X)$$

and the colimit runs over the covering sieves $\mathscr{R}(B)$ of $\overline{B}$. Since the open cover topology on $\mathscr{B}^{\text{op}}$ is subcanonical, $\overline{B} \in G = \text{Sh}(\mathscr{B}^{\text{op}})$.

For presheaves $X, Y \in Set^{\mathscr{B}}$, let us denote by $[X,Y]$ the set of natural transformations from X to Y. To prove that $\overline{W}$ is an atom in $\mathscr{G}$ for each Weil algebra W, it is enough to show

$$(aX)^{\overline{W}} \simeq a(X^{\overline{W}})$$

since, if F is a sheaf, we have the sequence of isomorphisms

$$\begin{aligned}[aX, F_{\overline{W}}] &\simeq [(aX)^{\overline{W}}, F] \simeq [a(X^{\overline{W}}), F] \\ &\simeq [X^{\overline{W}}, F] \simeq [X, F_{\overline{W}}] \simeq [aX, a(F_{\overline{W}})].\end{aligned}$$

In turn, the desired isomorphism follows from

$$(lX)^{\overline{W}} \simeq l(X^{\overline{W}}),$$

which is established as follows. We have

$$l(X^{\overline{W}})(B) = \text{colim}_{\mathscr{R}(B)} [\mathscr{R}, X^{\overline{W}}] = \text{colim}_{\mathscr{R}(B)} [\mathscr{R} \times \overline{W}, X]$$

and

$$l(X)^{\overline{W}}(B) = l(X)(W \otimes_\infty B) = \text{colim}_{\mathscr{R}'(\overline{W \otimes_\infty B})} [\mathscr{R}', X]$$

but every open cover of $\overline{W \otimes_\infty B}$ comes from one of $\overline{B}$ by pushout along the inclusion $B \hookrightarrow B \otimes_\infty W$, therefore there is a bijection between the two sorts of covers since W has a unique point. In fact, if $\mathscr{R}'(\overline{W \otimes_\infty B})$ corresponds to $\mathscr{R}(\overline{B})$ then $\mathscr{R}' = \mathscr{R} \times \overline{W}$.

Theorem 11.20 *[61]* **Postulate K** *is valid in* $(\mathscr{G},R)$.

Proof Let us first show that the statement

$$\forall \bar{x} \in R^n \; [\neg(\bigwedge_{i=1}^{n}(x_i = 0)) \Rightarrow \bigvee_{i=1}^{n}(x_i \in R^*)]$$

is valid in $\mathscr{G}$, where $R^* \subset R$ denotes the subobject of invertible elements of R. Let $B \in \mathscr{B}$ and let $\overline{B} \xrightarrow{x_l} R$ be such that

$$\vdash_{\overline{B}} \neg(\bigwedge_{i=1}^{n}(x_i = 0)).$$

Denote by $\bar{x_i} \in \mathscr{B}$ the actual elements of B corresponding to the points x_i of $\overline{B}$. Since every finitely generated ideal is germ determined, the quotient morphism

$$\beta : B \longrightarrow C = B/(\bar{x_1}, \ldots, \bar{x_n})$$

is in $\mathscr{B}$. Since under β the images of the points $\bar{x_i}$ are 0, C is covered by the empty family therefore $C = \{0\}$ and so $(\bar{x_1}, \ldots, \bar{x_n}) = B$. Consider the family

$$\{B \xrightarrow{\xi_i} B[x_i{}^{-1}] \mid i = i, \ldots, n\}.$$

If $p \colon B \longrightarrow \mathbb{R}$ is any point of B then the images of the $p(x_i)$ generate the unit ideal of $\mathbb{R}$ and so at least one of them, say $p(x_i)$ is invertible. Therefore for that particular i, p factors through ξ_i. This says that the family is covering for $\mathscr{B}^{\mathrm{op}}$. It follows from this that

$$\vdash_{\overline{B}} \bigvee_{i+1}^{n}(x_i \in R^*).$$

For the converse, we first show that for every $B \in \mathscr{B}$,

$$\vdash_{\overline{B}} \forall x \in R \, (x \in R^* \Rightarrow \neg(x = 0)).$$

If $x \colon \overline{B} \longrightarrow R$ satisfies $\vdash_{\overline{B}} (x \in R^*)$ and $\beta : B \longrightarrow C = B/(\bar{x_1}, \ldots, \bar{x_n})$ sends $\bar{x}$ to 0 then C must be the zero ring, which is a contradiction. Therefore

$$\vdash_{\overline{B}} \neg(x = 0),$$

and we have proved the implication

$$\vdash_{\overline{B}} \bigwedge_{i=1}^{n}(x_i \in R^*) \Rightarrow \neg(x_i = 0).$$

Use Exercise 1.16 in Heyting logic to complete the argument. □

It follows from Theorem 11.4, Theorem 11.17, Theorem 11.18, and Theorem 2.10, that $(\mathscr{G}, R)$ is a well adapted model of SDG. In particular, Postulate O holds in $(\mathscr{G}, R)$ as has already been observed in the previous chapter. In the rest of this section we establish the validity in $(\mathscr{G}, R)$ of the axioms of integration, including Postulate F.

Theorem 11.21 [99] **Axiom I** *is valid in* $(\mathscr{G}, R)$.

Proof Recall that this axiom says

$$\forall f \in R^{[0,1]}\, \exists! g \in R^{[0,1]}\, [g' = f \wedge g(0) = 0].$$

The lemma of Calderón-Quê-Reyes [99] says the following: let X, Y be closed subsets of $\mathbb{R}^n, \mathbb{R}^m$ respectively. Then, denoting by $\mathfrak{m}_X^\infty$ the ideal of flat functions on X, similarly for Y, there is an identity

$$\mathfrak{m}_{X \times Y}^\infty = \mathfrak{m}_X^\infty \cdot \pi_1 + \mathfrak{m}_Y^\infty \cdot \pi_2$$

where via the projections $\pi_1 \colon \mathbb{R}^n + \mathbb{R}^m \longrightarrow \mathbb{R}^n$, $\pi_2 \colon \mathbb{R}^n + \mathbb{R}^m \longrightarrow \mathbb{R}^m$, the ideals $\mathfrak{m}_X^\infty$ and $\mathfrak{m}_Y^\infty$ may be regarded as ideals of $C^\infty(\mathbb{R}^n \times \mathbb{R}^m)$. In particular, if Y is a closed subset of $\mathbb{R}^m$, then

$$\mathfrak{m}_Y^\infty \cdot \pi_2 = \mathfrak{m}_{\mathbb{R}^n \times Y}^\infty.$$

We now prove the statement of the theorem. Let $B \in \mathscr{B}$ of the form $B = C^\infty(\mathbb{R}^n)/I$ and let $f \in_{\overline{B}} R^{[0,1]}$. Such an f (see [61] for details) is represented by a smooth mapping $F \colon \mathbb{R}^n \times \mathbb{R}^m \longrightarrow \mathbb{R}$ modulo the ideal

$$K = \overline{I \circ \pi_1 \circ \mathfrak{m}_{[0,1]}^o \circ \pi_2},$$

the germ determined reflection of the ideal generated by $I \circ \pi_1$ and $\mathfrak{m}_{\|0,1\|}^o \circ \pi_2$ in $C^\infty(\mathbb{R}^{n+1})$.

Define

$$G(\bar{x}, t) = \int_0^t F(\bar{x}, u)du$$

for $t \in [0,1]$, and let g be the class of G modulo K. Clearly g satisfies the required conditions of $\int_0^t f(u)du = (t)$ provided it is well defined. In order to show this we need to establish that if $F \in K$ then also $G \in K$. That $F \in K$ says that locally, that is, on a covering of B in the site $\mathscr{B}^{\text{op}}$, that is, on some $\{B \longrightarrow B_\alpha\}_\alpha \in Cocov(\mathscr{B})$, with I_α the ideal of definition of B_α, F belongs to the ideal generated by $I \circ \pi_1$ and $\mathfrak{m}_{[0,1]}^o \circ \pi_2$. This says that for each α there exist $l_i^\alpha \in I_\alpha$ and $\tilde{l}_i^\alpha \in \mathfrak{m}_{[0,1]}^o$, as well as $\lambda_i^\alpha, \tilde{\lambda}_i^\alpha$ in $C^\infty(\mathbb{R}^n \times \mathbb{R})$ such that

$$\vdash_{\overline{B_\alpha}} F(\bar{x}, t) = \sum_i l_i^\alpha(\bar{x})\lambda_i^\alpha(\bar{x}, t) + \sum_j \tilde{l}_j^\alpha(t)\tilde{\lambda}_j^\alpha(\bar{x}, t).$$

Notice that the second summand vanishes on $\mathbb{R}^n \times [0,1]$ —denote it by $\sigma^\alpha(\bar{x},t)$. Integrating with respect to t now gives

$$\vdash_{\overline{B\alpha}} G(\bar{x},t) = \sum_i l_i^\alpha(\bar{x})\mu_i^\alpha(\bar{x},t) + \int_0^t \sigma^\alpha(\bar{x},u)du$$

and while the first summand belongs to $I \circ \pi_1$, the second still vanishes on $\mathbb{R}^n \times [0,1]$, that is, it belongs to $\mathfrak{m}^o_{\mathbb{R}^n \times [0,1]}$ which, by the lemma mentioned earlier is equal to $\mathfrak{m}^o_{[0,1]} \circ \pi_2$ since clearly $\mathfrak{m}^o_{[0,1]} = \mathfrak{m}^\infty_{[0,1]}$ and similarly for $\mathfrak{m}^o_{[0,1]} \times \mathbb{R}^n$. Hence, G is indeed in K. □

In what follows we shall need also a result referred to in [61] as the *Positivstellensatz*, adapted here to the category $\mathscr{B}$.

Lemma 11.22 *(Positivstellensatz)*
Let $q\colon C^\infty(\mathbb{R}^m) \twoheadrightarrow B$ be a presentation of $B \in \mathscr{B}$ with kernel J. Let $g \in C^\infty(\mathbb{R}^m)$ and let $\bar{g} = q(g) \in B$. Then the following are equivalent conditions:

(i) *g maps $Z(J)$ into $H = \{x \in \mathbb{R} \mid x \geq 0\}$.*
(ii) *$\vdash_{\overline{B}} 0 \leq \hat{F}$ where $\hat{g}\colon C^\infty(\mathbb{R}) \longrightarrow B$ is the homomorphism sending $\mathrm{id}_\mathbb{R}$ to $\bar{g}$.*

Proof Let B be as in the statement of the lemma and assume (i). To prove (ii) means to prove that $\hat{g}\colon C^\infty(\mathbb{R}^m) \longrightarrow B$ factors through $C^\infty(\mathbb{R}^m) \longrightarrow C^\infty(H)$, or that $\hat{g}$ annihilates the ideal I of functions vanishing on H. It is enough to show that $\hat{g}$ annihilates the ideal I' of functions vanishing on some open subset of $\mathbb{R}$ containing H. To this end we let $f \in I'$ vanishing on $U \subset \mathbb{R}$ open and containing H. Then

$$Z(J) \subset g^{-1}(H) \subset g^{-1}(U) \subset \mathbb{R}^m,$$

and $f \circ g$ vanishes on $g^{-1}(U)$. Since $g^{-1}(U)$ is open, the germ of $f \circ g$ at any $p \in Z(J)$ is zero. This implies that $\hat{g}(f)|_p \in B|_p$ is zero for any point p of B and since B is germ determined, $\hat{g}(f) = 0$.

Conversely, assume (ii) and let $p \in Z(J)$. Then $p\colon C^\infty(\mathbb{R}^M) \longrightarrow \mathbb{R}$ factors across B as (say) $\bar{p}\colon B \longrightarrow \mathbb{R}$. The composite

$$C^\infty(\mathbb{R}) \xrightarrow{\hat{g}} B \xrightarrow{\bar{p}} \mathbb{R}$$

is an element of R defined at stage $\overline{\mathbb{R}}$. Since $\vdash_{\overline{B}} 0 \leq \hat{g}$, we have $\vdash_{\overline{\mathbb{R}}} 0 \leq \hat{g}$. On the other hand, $g(p) < 0$ implies $\vdash_\mathbb{R} \hat{g} < 0$. This contradicts one of the items of Axiom O. Therefore $g(p) \notin H$ is incompatible with the assumption, hence (i) holds. □

Theorem 11.23 [8] **Axiom P** *is valid in* $(\mathcal{G}, R)$.

Proof This axiom is clearly implied by

$$\forall f \in R^{[0,1]}\ \forall t \in [0,1]\ \left[f(t) > 0 \Rightarrow \int_0^1 f(t)dt > 0\right]$$

so we show the above is valid in $\mathscr{G}$. Let f be defined at stage $\overline{B}$ for $B = C^\infty(\mathbb{R}^n)/I$ with I a germ determined ideal. That means that f is represented by a smooth mapping $F\colon \mathbb{R}^n \times \mathbb{R} \longrightarrow \mathbb{R}$ modulo I. With no loss of generality we may assume given $t \in_{\overline{B}} [0,1]$, itself represented by a smooth mapping $\gamma\colon \mathbb{R}^n \longrightarrow \mathbb{R}$ modulo I, such that $\gamma^*(m^o_{[0,1]}) \subset I$. The assumption on f translates into

$$F(-,\gamma(-))|_{Z(I)} > 0$$

where $Z(I) \subset \mathbb{R}^n$ is the closed set of the zeroes of the ideal I. To show:

$$\left.\int_0^1 F(-,u)du\right|_{Z(I)} > 0$$

on account of Lemma 11.22. In turn, it is enough to show that for each $\bar{x} \in Z(I)$,

$$F(\bar{x},-)|_{[0,1]} > 0.$$

For $t \in [0,1]$, the constant function γ with value t is smooth and such that $t^*[m^o_{[0,1]}] \subset I$. Therefore $F(\bar{x},t) > 0$ for every $\bar{x} \in Z(I), t \in [0,1]$, which finishes the proof. □

Theorem 11.24 [8] **Axiom X** *is valid in* $(\mathscr{G}, R)$.

Proof It is shown in [8] that the axiom of existence of flat functions, stated therein as

$$\vdash_1 \exists g \in R^R\ \forall t \in R\ [(t \leq 0 \Rightarrow g(t) = 0) \wedge (t > 0 \Rightarrow g(t) > 0)]$$

is valid in $\mathscr{G}$. We use it to prove the validity of Axiom X in $\mathscr{G}$, that is, the validity in $\mathscr{G}$ of the statement

$$\begin{aligned}\forall a,b \in R\ \big[a < b \Rightarrow \exists h \in R^R\ \forall t \in R\ \ [\,&(a < t < b) \Rightarrow h(t) > 0 \\ &\wedge (t < a \vee t > b) \Rightarrow h(t) = 0]\big].\end{aligned}$$

Let $a, b \in R$ be given at some stage $\overline{B}$ for $B \in \mathscr{B}$. Since $g \in R^R$ is globally given we may restrict it to the same stage $\overline{B}$. Set

$$h(t) = g(t-a)g(b-t).$$

If $a < t < b$, $t - a > 0$ and $t - b > 0$ hence $g(t-a) > 0$ so $h(t) > 0$. If $t < a$, $t - a < 0$ and so $h(t) = g(t-a) \cdot g(t-b) = 0$. Similarly if $t > b$ as then $b - t < 0$ and so $g(b-t) = 0$ hence $h(t) = 0$. Therefore, h is as required. □

Theorem 11.25 [27] **Axiom C** *is valid for* $(\mathcal{G}, R)$.

Proof Recall that this axiom says

$$\forall f \in R^{[0,1]}\ \forall t \in (0,1) f(t) > 0 \Rightarrow \exists a,b \in R\ \big[0 < a < t < b < 1 \\ \wedge \forall u \in (a,b)\ (f(u) > 0)\big].$$

In order to test its validity in $\mathcal{G}$, let

$$f \in_{\overline{B}} R^{[0,1]}$$

where $B = C^{\infty}(\mathbb{R}^n/I) \in \mathscr{B}$. We let f be represented by a smooth mapping

$$F\colon \mathbb{R}^n \times \mathbb{R} \longrightarrow \mathbb{R}$$

defined modulo K, with K the germ determined reflection of the ideal generated by $(I \circ \pi_1, m^o_{[0,1]} \circ \pi_2)$. Let $t \in_{\overline{B}} (0,1)$ and assume that

$$\vdash_{\overline{B}} f(t) > 0.$$

Using results of [61, 100], we have that such a t must be represented, modulo I, by a smooth mapping $\gamma\colon \mathbb{R}^n \longrightarrow \mathbb{R}$ and that the assumption above translates into

$$\forall \bar{x} \in Z(I)[\gamma(\bar{x}) \in (0,1) \subset \mathbb{R}^n]$$

and

$$\forall \bar{x} \in Z(I)[F(\bar{x}, \gamma(\bar{x})) > 0]$$

where as usual, $Z(I)$ is the set of zeroes of I.

By continuity of F one has, for each $\bar{x} \in Z(I)$, an open $U_{\bar{x}} \subset \mathbb{R}^n$ containing $\bar{x}$, as well as an open interval $(a,b) \subset \mathbb{R}$ containing $\gamma(\bar{x})$ such that

$$\forall (\bar{z}, y) \in (U_{\bar{x}} \cap Z(I)) \times (a,b)\,[F(\bar{z}, y) > 0].$$

Since $\gamma\colon \mathbb{R}^n \longrightarrow \mathbb{R}$ is also continuous, $\gamma^{-1}(a_{\bar{x}}, b_{\bar{x}})$ is open in $\mathbb{R}^n$ and contains $\bar{x}$.

Let

$$V_{\bar{x}} = U_{\bar{x}} \cap \gamma^{-1}(a_{\bar{x}}, b_{\bar{x}}).$$

$$\forall (\bar{z}, y) \in (V_{\bar{x}} \cap Z(I)) \times (a_{\bar{x}}, b_{\bar{x}})\ [F(\bar{z}, y) > 0] \qquad (*)$$

as well as

$$\forall \bar{z} \in V_{\bar{x}} \cap Z(I)\ [0 < a_{\bar{x}} < \gamma(\bar{z}) < b_{\bar{x}} < 1]. \qquad (**)$$

Now, the $\{V_{\bar{x}} \cap Z(I)\}$ form an open covering of $Z(I)$ in the induced topology of $Z(I) \subset \mathbb{R}^n$ which may be reduced to a countable subcovering $\{V_{\bar{x}_\alpha} \cap Z(I)\}_\alpha$. For each α, let B_α be given by the following pushout diagram in $\mathscr{B}$:

$$\begin{array}{ccc} C^\infty(\mathbb{R}^n) & \longrightarrow & C^\infty(V_{\bar{x}_\alpha}) \\ \downarrow & & \downarrow \\ B & \longrightarrow & B_\alpha \end{array}$$

In this way we get, by the definition of the Grothendieck topology on $\mathscr{B}^{\mathrm{op}}$, a cocovering $\{B \longrightarrow B_\alpha\}$ of B. Strictly speaking, a covering of $\mathbb{R}^n$ results from the $V_{\bar{x}_\alpha}$ together with the complement of $Z(I)$ but the latter gets eliminated when taking the pushout along $C^\infty(\mathbb{R}^n) \twoheadrightarrow B$.

From the choice of the $a_\alpha = a_{\bar{x}_\alpha}$ and $b_\alpha = b_{\bar{x}_\alpha}$, we get

$$\vdash_{\overline{B}} 0 < a_\alpha < c < b_\alpha < 1$$

as follows from (**) and the *Positivstellensatz* for germ determined ideals. (Indeed, if J_α is the ideal of definition of B_α, it is germ determined and $Z(J_\alpha) = V_{\bar{x}_\alpha} \cap Z(I)$). It remains to prove that

$$\vdash_{\overline{B_\alpha}} \forall u \in (a_\alpha, b_\alpha)\, f(u) > 0.$$

Let $v \in_{\overline{B_\alpha}} (a_\alpha, b_\alpha)$. This v will be represented modulo J_α by a smooth mapping $\psi \colon \mathbb{R}^n \longrightarrow \mathbb{R}$ which satisfies

$$\forall \bar{z} \in V_{\bar{x}_\alpha} \cap Z(I)\ \psi(\bar{z}) \in (a_\alpha, b_\alpha).$$

From (*) it follows that

$$\forall \bar{z} \in V_{\bar{x}_\alpha} \cap Z(I)\ F(\bar{z}, \psi(\bar{z})) > 0.$$

Therefore, by Lemma 11.22,

$$\vdash_{B_\alpha} f(v) > 0.$$

□

The following two lemmas are consequences of the axioms for SDG and Axiom I (integration axiom).

Lemma 11.26 (Hadamard's Lemma) *[90] The following holds in $\mathcal{G}$:*

$$\forall a, b \in R\, \forall f \in R^{[a,b]}\, \forall x, y \in [a,b] \left[f(y) - f(x) = (y - x) \cdot \int_0^1 f'(x + t(y-x))dt \right]$$

where $[a,b] = \{x \in R \mid a \le x \le b\}$.

Proof We use the integration axiom (Axiom I) in this proof. For $x, y \in [a,b]$ let $\varphi\colon [0,1] \longrightarrow [a,b]$ be the map $\varphi(t) = x + t(y-x)$ and compute

$$\begin{aligned} f(y) - f(x) &= f(\varphi(1) - \varphi(0)) \\ &= \int_0^1 (f \circ \varphi)'(t)dt \\ &= \int_0^1 (y-x)(f' \circ \varphi)(t)dt \\ &= (y-x)\int_0^1 f'(x + t(y-x))dt \end{aligned}$$

using the chain rule. □

Lemma 11.27 *[90] The following holds in $\mathcal{G}$:*

$$\forall f \in R^R\, [\forall x \in R\, (xf(x) = 0) \Rightarrow \forall x \in R\, (f(x) = 0)].$$

Proof Given the hypothesis we wish to show, for $\lambda \in R$, that $f(\lambda) = 0$. Let $\varphi_\lambda(x) = xf(\lambda x)$. Then, $\varphi_1(x) = xf(x) = 0$ and $\varphi_1(\lambda x) = f(\lambda x) + \lambda x f(\lambda x) = \frac{\partial}{\partial x}\varphi_\lambda(x) = 0$ for all x since $\varphi_1(\lambda x)\lambda x f(\lambda x) = 0$. Therefore $\frac{\partial}{\partial x}\varphi_\lambda(x) = 0$ for all x. By the integration axiom (Axiom I), $\varphi_\lambda(x) = \varphi_\lambda(0) = 0$ for all x. In particular, $\varphi_\lambda(1) = f(\lambda) = 0$. □

Theorem 11.28 [90] *Postulate F is valid for* $(\mathcal{G}, R)$.

Proof Postulate F holds in $\mathcal{G}$. For uniqueness use Lemma 11.27. For existence use Hadamard's Lemma (Lemma 11.26). □

12
$\mathscr{G}$ as a Model of SDT

This last chapter is devoted to establishing that $\mathscr{G}$ is a model of SDT. Although there are other known well adapted models to SDG [89], the topos $\mathscr{G}$ is so far the only one which, to our knowledge and as shown here, is also a model of SDT. With the exception of the validity of the uniqueness part in Postulate S, which is new, all other proofs of validity in $\mathscr{G}$ of axioms and postulates of SDG/SDT included here are collected from various sources, among them [8, 20, 26, 27, 35, 44, 96, 99], and are so indicated in the text.

12.1 Validity in $\mathscr{G}$ of the Basic Axioms of SDT

We already know from Theorem 11.4 that ι preserves transversal pullbacks, 1, and open coverings, that is, that $(\mathscr{G}, R)$ is a well adapted model of ringed toposes with $R = \iota(\mathbb{R}) = \overline{C^\infty(\mathbb{R})}$. By Theorem 9.17, $R \in \mathscr{G}$ is an Archimedean (local) ordered ring in $\mathscr{E}$, so that **Postulate O** holds for $(\mathscr{G}, R)$.

That **Postulate E** also holds in $(\mathscr{G}, R)$ is a consequence of general facts about the site $\mathscr{B}$ of definition [25, Appendix], taking into account also the following two facts about $\mathscr{B}$.

For A an object of $\mathscr{B}$, say $A = C^\infty(\mathbb{R}^n)/I$ where I is a germ determined ideal, $\operatorname{Spec}(A)$ denotes $Z(I)$, the zeroes of I.

Lemma 12.1 [96] *For any A an object of $\mathscr{B}$,* $\operatorname{Spec}(A)$ *satisfies the covering principle.*

Lemma 12.2 [25] *For any A an object of $\mathscr{B}$,* $\operatorname{Spec}(A) \subset E(A)$.

In the topos $\mathscr{G}$, not only the algebraic infinitesimals $D_r(n)$ are representable but so are the logical infinitesimals, such as $\Delta(n) \subset R^n$, as already shown in Proposition 11.5 that it is represented by $C_0^\infty(\mathbb{R}^n)$ which is an object of $\mathscr{B}$.

Before proving the validity of Axiom G in $\mathscr{G}$, we recall from Section 6.1 how the object $C_0^g(R^n,R)$ of germs is internally defined in any model $(\mathscr{E},R)$ of SDG. Denote by $\mathrm{Partial}(R^n,R)$ the subobject of Ω^{R^n} whose objects are the partial maps defined as usual but with respect to the intrinsic topological structure. Denote by $\partial\colon \mathrm{Partial}(R^n,R) \longrightarrow \Omega^{R^n}$ the functor which assigns to a partial map f its domain $\partial(f)$. A germ at 0 is an equivalence class of elements $f \in \mathrm{Partial}(R^n,R)$ such that $0 \in \partial(f) \in P(R^n)$. The equivalence relation for partial maps f and g is given by

$$f \sim g \Leftrightarrow \exists U \in P(R^n)\,[0 \in U \subset \partial(f) \cap \partial(g) \wedge f|_U = g|_U].$$

The quotient of $C^\infty(R^n,R)$ by this equivalence relation is one of the ways to define $C_0^g(R^n,R)$.

Theorem 12.3 *(***Axiom G***)* [35] *The restriction map*

$$j\colon C_0^g(R^n,R) \longrightarrow R^{\Delta(n)}$$

is invertible in $\mathscr{G}$.

Proof Surjectivity is a consequence of a stronger result, to wit, that in $\mathscr{G}$, germs are globally defined, that is, the restriction map $R^{R^n} \longrightarrow R^{\Delta(n)}$ is an epimorphism. That this is the case is itself a consequence of Proposition 11.5.

Injectivity uses Proposition 11.5, Proposition 11.12 and Corollary 5.20. Let $f,g\colon \overline{A} \longrightarrow \mathrm{Partial}(R^n,R)$, where $A = C^\infty(\mathbb{R}^s)/I$ is an object of $\mathscr{B}$. We may assume without loss of generality that f and g have the same domain H. This means that the corresponding morphisms $f,g\colon H \longrightarrow R$ where $H\colon \overline{A} \longrightarrow \Omega^{R^n}$ and $(0,\mathrm{id})\colon \overline{A} \longrightarrow R^n \times \overline{A}$ factors through $H \subset R^n \times \overline{A}$. Now, this does not say that $H \subset R^n \times \overline{A}$ is intrinsic open.

However, by the covering principle (Postulate E) and so its consequence Corollary 5.20, there is an intrinsic open $G \subset R^n \times \overline{A}$ such that

$$\{0\} \times \overline{A} \subset G \subset H \subset R^n \times \overline{A}.$$

Now, by Proposition 11.13 there is $W \subset \mathbb{R}^n \times \mathbb{R}^s$ such that $G = \iota W \cap \overline{A}$. That is, $G = C^\infty(W)/I \mid W$. We now use Proposition 11.5. Recall that J denotes the ideal of functions whose germ at 0 is null. Now, $f,g\colon G \longrightarrow R$ are represented in $C^\infty(W)$ and coincide on $\Delta(n) \times \overline{A} \subset \iota W$. This means that $(f-g) \in^o (J,I)|_W$, in the sense that for each point $(0,p) \in Z(J,I) \subset W$, there is a neighbourhood V_p where

$$f - g = \sum \varphi_i h_i$$

where $\varphi_i \in C^\infty(V_p)$ and $h_i \in I$, since the part corresponding to J is 0. So, there is an open $V \subset W$ such that $\{0\} \times Z(I) = Z(J,I) \subset V$ for which $(f-g) \in^o I|_V$. In other words, f and g are equal on $U = \iota V \cap \overline{A} \subset R^n \times \overline{A}$. Since $U \in P(R^n) \times \overline{A}$, we have, with this U, the desired factorization. □

Theorem 12.4 (Axiom M) *For any $n > 0$, the object $\Delta(n) = \neg\neg\{0\}$ of $\mathscr{G}$, where $0 \in R^n$, is an atom in $\mathscr{G}$.*

Proof The proof is similar to that of Theorem 11.18 since $\Delta(n) = \neg\neg\{0\}$ is representable in $\mathscr{G}$ (Proposition 11.5). □

The central theme of [96] is to state and prove a theorem of local inversion which would explain the need for Grothendieck to introduce the étale topos. Of the various equivalent versions of it, the following is established in [95] for model $(\mathscr{E}, R)$ other than the Dubuc topos but the proof applies to the latter for the same reasons and it is given below.

Theorem 12.5 (Postulate I.I) *For positive integer n,*

$$\forall f \in \Delta(n)^{\Delta(n)} \left[(f(0) = 0 \wedge \text{Rank}(D_0 f) = n) \Rightarrow f \in \text{Iso}\left(\Delta(n)^{\Delta(n)}\right) \right]$$

holds in $(\mathscr{G}, R)$.

Proof Denote by $\text{GL}(n)$ the subobject of $\mathfrak{M}_{m\times m}(R)$ consisting of the invertible matrices. Implicit in the proof of Proposition 2.34 is the equivalence

$$\text{Rank}(A) = n \Leftrightarrow A \in \text{GL}(n)$$

established in [59] for any $A \in \mathfrak{M}_{m\times m}(R)$. We wish to show then, equivalently, the following:

For positive integer n,

$$\forall f \in \Delta(n)^{\Delta(n)} \left[(f(0) = 0 \wedge D_0 f \in \text{GL}(n)) \Rightarrow f \in \text{Iso}\left(\Delta(n)^{\Delta(n)}\right) \right].$$

Given the hypothesis of the theorem, it follows from Postulate F that

$$\exists \varphi \colon R^n \times R^n \to R^{n^2} \, \forall x, x' \in R \left[f_i(x') - f_i(x) = \sum_j \varphi_{ij}(x, x')(x'_i - x_i) \right].$$

We thus already have

$$\forall x, x' \in R \left[\varphi(x, x') \in \text{GL}(n) \Rightarrow (f(x') = f(x) \Rightarrow x' = x) \right].$$

Using now Postulate K, we are reduced to proving the validity of the following two formulas (where we have omitted the obvious universal quantifiers at the outside):

$$\left(\bigvee_{i=1}^{n} (x'_i - x_i) \# 0 \right) \vee \left(\bigvee_{i=1}^{n} (x''_i - x_i) \# 0 \right) \vee \left(\varphi(x', x'') \in \text{GL}(n) \right)$$

and

$$\left(\bigvee_{i=1}^{n}(y_i - f_i(x))\#0\right) \vee \left(\exists x' \left(f(x') = y \wedge \varphi(x,x') \in \mathrm{GL}(n)\right)\right).$$

These formulas are coherent and, since the hypothesis may also be rendered coherent, it is enough to prove their validity in *Set* [80], which is indeed the case. □

In [25], a certain Postulate WA2 on the existence and uniqueness of solutions to ordinary differential equations was introduced in the context of SDG. Before dealing with Postulate S here, we remind the reader of it and give a proof of its validity in the topos $\mathscr{G}$. The existence part, which was merely indicated in Theorem 19 of [35], is given a more explicit form here. As for the uniqueness, which is actually needed for the flow condition in the existence part, we give a proof that differs from that of [35], which was incorrect. The error in it was found by M. Makkai and communicated to us by G. E. Reyes [101].

Although Postulate WA2 was originally stated in [25] for germs whose domain is some $M = R^m$, the version we gave of it in Chapter 6, the assumption was made that $M \subset R^m$ is an intrinsic open containing 0. Nevertheless, in the proofs of validity in $\mathscr{G}$ given below, we simplify the set-up by letting $M = R^m$. The reader is asked to carry out the proofs in the more general case using for this the representability of intrinsic opens of R^m in $\mathscr{G}$.

Theorem 12.6 (Postulate WA2) *The statement*

$$\forall g \in R^{mR^m}\, \exists! f \in R^{mR^m \times \Delta} \forall x \in R^m \forall t \in \Delta \left[f(x,0) = x \wedge \frac{\partial f}{\partial t}(x,t) = g(f(x,t)) \right]$$

is valid in $(\mathscr{G}, R)$.

Proof The existence part of the statement follows from the classical theory of ODE. It may be given as follows. Let $g \in_{\overline{A}} R^{mR^m}$ for $A = C^\infty(\mathbb{R}^n)/I$ where $I \subset C^\infty(\mathbb{R}^n)$ is any local ideal. This means that g is represented by a smooth mapping

$$G\colon \mathbb{R}^n \times \mathbb{R}^m \longrightarrow \mathbb{R}^m$$

defined modulo $(I,0)$, the local ideal generated by I and 0. By the classical theory of differential equations, there is a smooth

$$F\colon U \longrightarrow \mathbb{R}^m$$

defined on U open in $\mathbb{R}^n \times \mathbb{R}^m \times \mathbb{R}$ such that $\mathbb{R}^n \times \mathbb{R}^m \times \{0\} \subset U$ and which is a solution of the differential equation determined by G, that is, so that it satisfies

$$\frac{\partial F}{\partial t}(x,t) = G(F(x,t))$$

and

$$F(x,0) = x.$$

Such an F is defined modulo $(I,0,J)$. From any such F we get the desired solution $f \in_{\overline{A}} R^{mR^m \times \Delta}$.

The uniqueness of solutions can now be argued as follows. Assume that $f \in_{\overline{A}} R^{mR^m \times \Delta}$ and $l \in_{\overline{A}} R^{mR^m \times \Delta}$ are both solutions to the differential equation associated with the given $g \in_{\overline{A}} R^{mR^m}$, both defined (w.l.o.g.) at the same stage $\overline{A}$, where $A = C^\infty(\mathbb{R}^n)/I$ and $I \subset C^\infty(\mathbb{R}^n)$ is any local ideal.

As above, let $G\colon \mathbb{R}^n \times \mathbb{R}^m \longrightarrow \mathbb{R}^m$ be a smooth mapping representing the given g at stage $\overline{A}$. Let $F, L\colon U \longrightarrow \mathbb{R}^m$ be smooth mappings representing f, l,wlog defined on the same open $U \subset \mathbb{R}^n \times \mathbb{R}^m \times \mathbb{R}$ and also defined at stage $\overline{A}$. We wish to show that

$$(F - L) \in^o (I,0,J)|_U.$$

We have that $Z(I,0,J) = Z(I) \times \mathbb{R}^m \times \{0\}$. Thus, we need to show that for any $x_0 \in \mathbb{R}^m$, $\lambda_0 \in Z(I)$, there exists an open $W \subset U$ such that $(\lambda_0, x_0, 0) \in W$ with

$$(F - L)|_W \in (I,0,J)|_W.$$

To this end, we proceed just as in Lemma 20 of [34], with a minor modification.

Since $0 \in (I,0)$, we can express it as finite sum

$$0 = \sum_{i=1}^{r} s_i(\lambda) h_i(\lambda, x),$$

where the $s_i(\lambda) \in I$ and the $h_i(\lambda, x) \in C^\infty(\mathbb{R}^n \times \mathbb{R}^m)$ are such that $(\lambda, x) \in Z(I,0) = Z(I)$.

Let $\mathbb{R}^r$ be considered as a parameter space, and let

$$\varphi\colon \mathbb{R}^r \times \mathbb{R}^n \times \mathbb{R}^m \rightarrow \mathbb{R}^m$$

be given by

$$\varphi(s, \lambda, x) = \sum_{i=1}^{r} s_i(\lambda) h_i(\lambda, x).$$

By the classical theory of ODE there exists $H \subset \mathbb{R}^n \times \mathbb{R}^m$ open, $(\lambda_0, x_0) \in H$, $V \subset \mathbb{R}^r$ open, $0 \in V$, $\varepsilon > 0$ and $\psi\colon V \times H \times (-\varepsilon, \varepsilon) \rightarrow \mathbb{R}^n$ so that for all

$(s,\lambda,x,t) \in V \times H \times (-\varepsilon,\varepsilon)$, we have $(\lambda, \psi(s,\lambda,x,t)) \in \mathbb{R}^n \times \mathbb{R}^m$ and such that

$$(\psi(s,\lambda,x,0) = x) \wedge (\frac{\partial \psi}{\partial t}(s,\lambda,x,t) = \varphi(s,\lambda,\psi(s,\lambda,x,t))).$$

There exist (by local Hadamard's Lemma) $k_i\colon V \times H \times (-\varepsilon,\varepsilon) \to \mathbb{R}^m$ such that

$$\psi(s,\lambda,x,t) = \psi(0,\lambda,x,t) + \sum_{i=1}^{r} s_i k_i(s,\lambda,x,t).$$

Since $\lambda_0 \in Z(I)$, and the $s_i \in I$, we have that $s_i(\lambda_0) = 0$. Let $H' \subset H$ be small enough so that $s_i(\lambda) \in V$ for all $\lambda \in H'$. It follows that

$$\psi(s(\lambda),\lambda,x,t) = \psi(0,\lambda,x,t) + \sum_{i=1}^{r} s_i(\lambda) k_i(s(\lambda),\lambda,x,t)$$

for all $(\lambda,x,t) \in W = (U \cap H') \times (-\varepsilon,\varepsilon)$.

By the uniqueness of solutions of classical ODE (and routine verification) it follows that for all $(\lambda,x,t) \in W$,

$$F(\lambda,x,t) = \psi(0,\lambda,x,t)$$

and

$$L(\lambda,x,t) = \psi(s(\lambda),\lambda,x,t).$$

Therefore

$$(F-L)|_W = \psi(0,\lambda,x,t) - \psi(s(\lambda),\lambda,x,t) = \sum_{i=1}^{r} s_i(\lambda) k_i(s(\lambda),\lambda,x,t)$$

and this concludes the proof modulo the routine verifications done below.

For all $(\lambda,x,t) \in W$,

(i) • $\begin{cases} F(\lambda,x,0) = x \\ \frac{\partial F}{\partial t}(\lambda,x,t) = G(\lambda,F(\lambda,x,t)) \end{cases}$

• $\begin{cases} \psi(0,\lambda,x,0) = x \\ \frac{\partial \psi}{\partial t}(0,\lambda,x,t) = \varphi(0,\lambda,\psi(0,\lambda,x,t)) = G(\lambda,\psi(0,\lambda,x,t)) \end{cases}$

(ii) • $\begin{cases} L(\lambda,x,0) = x \\ \frac{\partial L}{\partial t}(\lambda,p,t) = G(\lambda,L(\lambda,x,t)) \end{cases}$

• $\begin{cases} \psi(s(\lambda),\lambda,x,0) = x \\ \frac{\partial \psi}{\partial t}(s(\lambda),\lambda,x,t) = \varphi(s(\lambda),\lambda,\psi(s(\lambda),\lambda,x,t)) = G(\lambda,\psi(s(\lambda),\lambda,x,t)) \end{cases}$

□

Theorem 12.7 *(Postulate S) Let $(\mathcal{E},R)$ be a basic model of SDT. Let $m>0$. Then the statement*

$$\forall g \in R^{mR^m\times[0,1]} \exists! f \in R^m \times [0,1]^{R^m\times[0,1]\times\Delta} \forall x \in R^m \forall s \in [0,1] \forall t \in \Delta$$
$$\left[f(x,s,0)=(x,s) \wedge \frac{\partial f}{\partial t}(x,s,t)=g(f(x,s,t))\right]$$

is valid in $(\mathcal{G},R)$.

Proof The proof given above for the validity of Postulate 12.6 in $\mathcal{G}$ can easily be adapted to give one of the validity of Postulate S in $\mathcal{G}$, as follows. In the case of Postulate S, the R^m in the domain of g in Postulate WA2 is replaced here by $R^m \times [0,1]$. The modifications required in the proof of Theorem 12.6 are then the following. The local ideal $(I,0)$ generated by I and 0 must be replaced by the local ideal $(I,0,K)$ generated by I, 0 and $K=m^g_{[0,1]}$.

As for $(I,0,J)$, it is here replaced by $(I,0,K,J)$. Notice that, whereas

$$Z(I,0,J)=Z(I)\times\mathbb{R}\times\{0\}$$

in the case of Theorem 12.6, in that of the validity of Postulate S in $\mathcal{G}$, we must use that $Z(I,0,K,J)=Z(I)\times\mathbb{R}\times[0,1]\times\{0\}$. The details of the proof are otherwise the same and are left to the reader. □

12.2 Validity in $\mathcal{G}$ of the Special Postulates of SDT

We are then left with verifying the validity in $(\mathcal{G},R)$ of Postulate D as well as that of Postulate PT. We begin with the latter, which is the internal version of the preparation theorem in differential topology, and which was used to obtain the first of the two proofs of Mather's theorem (from [44, 26]) in Chapter 8. Notice, however, that the second proof of Mather's theorem (from [103]) also given in Chapter 8 does not need to resort to this axiom.

Theorem 12.8 (Postulate PT) *The following holds in $\mathcal{G}$ with $R=\overline{C^\infty(\mathbb{R})}$. Let $f\in R^{\Delta(n)}$ and let $f\in V\subset R^{\Delta(n)}$ be a weak-open neighbourhood of f. Let*

$$\Phi\colon V \longrightarrow V^{[0,1]}$$

be any morphism such that $\Phi(f)(s)=f$ for all $s\in[0,1]$. Then, if $(d\gamma)_{\Phi(f)}$ is surjective at

$$(\pi_{R^n}\colon [0,1]\times R^n \longrightarrow R^n, \pi_R\colon [0,1] \longrightarrow R),$$

it follows that $(d\gamma)_{\Phi|_{V'}}$ is surjective at

$$(\pi_{R^n}\colon V'\times[0,1]\times R^n \longrightarrow R^n, \pi_R\colon V'\times[0,1] \longrightarrow R)$$

for some weak-open neighbourhood V' so that $f \in V' \subset V$.

Proof Let $f \in_{\overline{A}} R^{\Delta(n)}$ be infinitesimally stable, where A is represented by $C^\infty(\mathbb{R}^r)/I$, and let $\Phi \in_{\overline{A}} R^{\Delta(n)^{R^{\Delta(n)} \times [0,1]}}$ be so that $\Phi(f,s) = f$ for every $s \in [0,1]$.

Applying the global sections functor, we get a mapping $F = \Gamma(\Phi)$,

$$F\colon Z(I) \times C^\infty{}_{\{0\}}(\mathbb{R}^n) \longrightarrow C^\infty{}_{\{0\}\times[0,1]}(\mathbb{R}^n \times [0,1])$$

which is smooth in the first variable, regarding $Z(I) = \Gamma(\overline{A})$ as a submanifold, and continuous in the second variable, regarding $C^\infty{}_{\{0\}}(\mathbb{R}^n) = \Gamma(R^{\Delta(n)})$ and similarly $C^\infty{}_{\{0\}\times[0,1]}(\mathbb{R}^n \times [0,1]) = \Gamma(R^{\Delta(n)} \times [0,1])$ endowed with the weak C^∞-topology.

The condition $\Phi(f,s) = f$ translates into $F(\lambda, f(\lambda))(s) = f(\lambda)$ for each $\lambda \in Z(I)$. Moreover, $f(\lambda)$ is infinitesimally stable and therefore $\alpha_{F(\lambda,f(\lambda))} \oplus \beta_{F(\lambda,f(\lambda))}$ is surjective. By [97] (Lemma 2.3) there exists some open V_λ in $Z(I) \times C^\infty{}_0(\mathbb{R}^n)$ such that $\alpha_{F|_{V_\lambda}} \oplus \beta_{F|_{V_\lambda}}$ is surjective, for each $\lambda \in Z(I)$.

We may assume that $V_\lambda = (U_\lambda \cap Z(I)) \times W_\lambda$ for some $U_\lambda \subset \mathbb{R}^r$ open in the usual sense, and $W_\lambda \subset C^\infty_{\{0\}}(\mathbb{R}^n)$ open in the (quotient) weak C^∞-topology. We can also restrict ourselves to considering a countable family

$$\{U_\alpha\} \subset \{U_\lambda\}$$

such that $\{U_\alpha \cap Z(I)\}$ covers $Z(I)$. Surjectivity of $\alpha_{F|_{V_\lambda}} \oplus \beta_{F|_{V_\lambda}}$ (at the corresponding projections) gives, for the representable objects $A_\alpha = C^\infty(U_\alpha)/(I/U_\alpha)$, that

$$\models_{\overline{A_\alpha}} \alpha_{\Phi|_{\Lambda(W_\lambda)}} \oplus \beta_{\Phi|_{\Lambda(W_\lambda)}} \text{ surjective.}$$

Now, $\Lambda(W_\alpha) \subset R^{\Delta(n)}$ is weak open, and the $\{\overline{A_\alpha} \longrightarrow \overline{A}\}$ form a cover. Therefore, the statement

$$\models \exists V \in W(R^{\Delta(n)})[f \in V \wedge \alpha_{\Phi|_V} \oplus \beta_{\Phi|_V} \text{ surjective}]$$

holds in $\mathscr{G}$ as desired. □

Our next task is to establish the validity of Postulate D in our test model $(\mathscr{G}, R)$. In the classical context a theorem of Sard's to the effect that the set of critical values of a smooth mapping has measure zero is used in order to derive several density results [47]. However, what is actually used is the fact that in every non-empty interval there are regular values, which is a statement equivalent to Sard's theorem within Boolean logic.

In our context, which is that of a topos, the internal logic is Heyting and both results cannot be proven to be equivalent. However, we will show that, when we restrict to functions defined on a logical infinitesimal domain, the positive version follows from the negative one. This is meaningful (only) in our test model $(\mathscr{G},R)$ where Axiom G holds. We begin therefore to establish Sard's theorem in a form that is meaningful in our context.

Theorem 12.9 (Sard's theorem) *The following statement is valid in $(\mathscr{G},R)$.*

$$\forall f \in R^{pR^n}\ \forall U \in P(R^p)\ [\neg\forall y \in R^p\ (y \in U \Rightarrow y \in \mathrm{Crit}(f))].$$

Proof Let $f_{\overline{A}} \in R^{pR^n}$ be represented by $F\colon \mathbb{R}^r \times \mathbb{R}^n \longrightarrow \mathbb{R}^p$, a smooth mapping defined modulo $I \circ \pi_1$, where $A = \mathscr{C}^\infty(\mathbb{R}^r)/I$. For our purposes it is certainly enough to suppose $U = (a,b)^p$ for $a,b \in R$ such that $\models_{\overline{A}} a < b$. Thus, we need to show that

$$\models_{\overline{A}} \neg\forall y \in R^p\ [y \in (a,b)^p \Rightarrow y \in \mathrm{Crit}(f)].$$

If $a,b \in_{\overline{A}} R$ are represented by $\alpha,\beta\colon \mathbb{R}^r \longrightarrow \mathbb{R}$ smooth mappings defined modulo J, then $\models_{\overline{A}} a < b$ if and only if $\forall t \in Z(J)\ (\alpha(t) < \beta(t))$.

Assume, for $B = C^\infty(\mathbb{R}^s)/J$, $\delta\colon \overline{B} \longrightarrow \overline{A}$ in $\mathscr{G}$ is such that

$$\models_{\overline{B}} \forall y \in R^p\ [y \in (a,b)^p \Rightarrow y \in \mathrm{Crit}(f)].$$

We need to show that $\overline{B} = 0$. If not, by the 'Nullstellensatz' (Proposition 11.8) $Z(J) \neq \emptyset$. Let $t_0 \in Z(J)$. Then

$$\alpha^\sharp(t_0) < \beta^\sharp(t_0)$$

where $\alpha^\sharp$ and $\beta^\sharp$ are those induced by α and β by the change of stage δ. Take any $z \in (\alpha^\sharp(t_0), \beta^\sharp(t_0)) \subset \mathbb{R}$. Then, there exists $\lambda \in \mathbb{R}$ with

$$z = \lambda \cdot \alpha^\sharp(t_0) + (1-\lambda)\cdot \beta^\sharp(t_0).$$

Consider the equivalence class modulo J of

$$\xi : \underline{x} \in \mathbb{R}^s \mapsto \lambda \cdot \alpha^\sharp(\underline{x}) + (1-\lambda)\cdot \beta^\sharp(\underline{x}) \in \mathbb{R}.$$

It defines an element $c \in_{\overline{B}} (a,b)$ and therefore

$$\models_{\overline{B}} c \in \mathrm{Crit}(f).$$

This amounts to

$$\models_{\overline{B}} \exists x \in R^n \left[f(x) = c \wedge \bigwedge_{H \in \binom{n}{m}} \det(D_x f)_H = 0 \right]$$

which means that there exists a covering

$$\{\overline{B_\alpha} \longrightarrow \overline{B}\}_\alpha$$

and

$$\xi_\alpha : \mathbb{R}^{s_\alpha} \longrightarrow \mathbb{R}^n$$

whose classes modulo the ideals (of definition of the $\overline{B_\alpha}$) J_α satisfy

$$\forall t \in Z(J_\alpha)\ (F(t,0) = \xi_\alpha(t))$$

and

$$\forall t \in Z(J_\alpha) \text{ every subset of } \left\{ \frac{\partial F}{\partial x_i}(t,0), \ldots, \frac{\partial F}{\partial x_n}(t,0) \right\} \text{ consisting of } p \text{ vectors}$$

has zero determinant.

Now, since $Z(J) = \bigcup_\alpha Z(J_\alpha)$, there must exist some a_α so that $t_0 \in Z(J_{\alpha_0})$. Considering the map $F_0 = F(t_0, -) \colon \mathbb{R}^n \longrightarrow \mathbb{R}^p$, F_0 is smooth and z is a critical value of F_0, but z is any point of the interval $(\alpha^\sharp(t_0), \beta^\sharp(t_0))$, and that contradicts classical Sard's theorem. Thus, we must have $\overline{B} = 0$. □

Theorem 12.10 (Postulate D) *The statement*

$$\forall f \in R^{p\Delta(n)}\ \exists y \in U\ [y \text{ is a regular value of } f]$$

is valid in $(\mathscr{G}, R)$ *for any* $n, p > 0$ *and* $U \in P(R^p)$.

Proof The statement to be proved holds in $(\mathscr{G}, R)$ may be equivalently stated as follows:

$$\forall f \in R^{p\Delta(n)}\ \exists y \in U\ \forall x \in \Delta(n) \left[\neg(f(x) = y) \vee \bigvee_{H \in \binom{n}{p}} \neg(\det(D_x f)_H = 0) \right].$$

For a given $f \in {}_{\overline{A}}R^{p\Delta(n)}$, consider the map $\Phi \in {}_{\overline{A}}R^{p(\Delta(n)\times U)}$ defined (implicitly at stage $\overline{A}$) as follows:

$$\Phi(x,y) = (f(x) - y)^2 + \sum (\det(D_x f)_H)^2.$$

Clearly, a sufficient condition for y to be a regular value of f is, for $y \in U$, that

$$\forall x \in \Delta(n)\ \neg(\Phi(x,y) = 0)$$

holds, and a necessary condition for y to be a critical value of f is, for $y \in U$ that

$$\exists x \in \Delta(n)\ (\Phi(x,y) = 0)$$

holds.

Therefore, it suffices for our purposes to establish that the former implies the latter, as the former is a consequence of Theorem 12.9 and the latter holds in $(\mathscr{G}, R)$.

From the former we derive, using the rules of Heyting logic, the following:

$$\models_{\overline{A}} \forall f \in R^{p\Delta(n)} \neg\exists x \in \Delta(n) \, \forall y \in U \, [\Phi(x,y) = 0]$$

which is equivalent intuitionistically [39, p. 29] to:

$$\models_{\overline{A}} \forall f \in R^{p\Delta(n)} \, \forall x \in \Delta(n) \, \neg\forall y \in U \, [\Phi(x,y) = 0].$$

Since U is of the form $\iota(V)$ for some $V \subset \mathbb{R}^p$, U is point determined [62]. For these objects, a sort of Markov principle is available and allows us to derive from the above the following:

$$\models_{\overline{A}} \forall f \in R^{p\Delta(n)} \, \exists \mathscr{U} \in \Omega^{(\Omega^{\Delta(n)})} \, [\, \mathscr{U} \text{ open cover of } \Delta(n) \\ \wedge \forall V \in \mathscr{U} \, \exists g \in U^V \, \forall x \in V \, \neg(\Phi(x,y) = 0)\,].$$

However, the intrinsic topological structure of $\Delta(n)$ is trivial (any intrinsic open object must contain the infinitesimal monad of each of its elements) and, given that $0 \in \Delta(n)$ and $\mathscr{U}$ is a covering, we must have $\Delta(n) = V$ for some $V \in \mathscr{U}$. In particular, we have

$$\models_{\overline{A}} \forall f \in R^{p\Delta(n)} \, \exists g \in U^{\Delta(n)} \, \forall x \in \Delta(n) \, \neg[\Phi(x,y) = 0].$$

To finish the proof we use the explicit description of $\Delta(n)$ in $\mathscr{G}$. First of all, for a given $f \in_{\overline{A}} R^{p\Delta(n)}$ the above gives the existence of an open covering of A in the site, that is, some covering $(A_i \twoheadrightarrow A)_{i \in I}$ such that for each $i \in I$ there is a $g_i \in_{\overline{A}} R^{p\Delta(n)}$ for which

$$\models_{\overline{A_i}} \forall x \in \Delta(n) \, \neg[\Phi(x, g_i(x)) = 0].$$

Finally, since R is a local ring, from the definition of Φ we get the formulation

$$\models_{\overline{A_i}} \forall x \in \Delta(n) \Big[\neg(f(x) = g_i(x)) \vee \bigvee_{H \in \binom{n}{m}} \neg(\det(D_x f)_H = 0) \Big]$$

which gives, for any $x \in_{\overline{A}} \Delta(n)$, $\models_{\overline{A_i}} \neg(f(x) = g_i(x))$ or

$$\models_{\overline{A_i}} \bigvee_{H \in \binom{n}{m}} \neg(\det(D_x f)_H = 0).$$

The second option does not depend on g_i. As for the first one, the assertion is equivalent to

$$\models_{\overline{A_i}} \neg(f(x) = g_i(0))$$

because of the monotonicity of $\neg\neg$ (remembering that $\Delta(n) = \neg\neg\{0\}$) that guarantees, since $\neg\neg(x = 0)$, that $\neg\neg(f(x) = f(0))$ and $\neg\neg(g_i(x) = g_i(0))$. This, for the element $c_i = g_i(0)$ at stage $\overline{A_i}$, gives

$$\models_{\overline{A_i}} \neg(f(x) = c)$$

and since the A_i form a covering of A, we get that

$$\models_{\overline{A}} \exists y \in U \; \forall x \in \Delta(n) \; \neg(\Phi(x,y) = 0).$$

□

References

[1] W. Ambrose, R. S. Palais, and I. M. Singer, Sprays, *Anais da Acad. Bras. de Ciências* **32-2** (1960) 163–178.

[2] V. I. Arnold, Normal forms of functions in the neighbourhood of degenerate critical points, *Uspekhi Matematicheskykh Nauk* **29** (1974) 11–49, *Russian Mathematical Surveys* **29** 19–48.

[3] V. I. Arnold, Singularity Theory, in: Jean-Paul Pier (ed.), *Development of Mathematics: 1950 - 2000*, Birkhäuser, Basel-Boston-New York (2000) 127–152.

[4] M. Artin, A. Grothendieck, J. L. Verdier, *Théorie des Topos et Cohomologie Étale des Schémas Lect. Notes in Math.* **269** Exposé IV. Springer-Verlag, Berlin, 1972.

[5] M. Barr, Toposes without points, *J. Pure Appl. Algebra* **5** (1974) 265–280.

[6] M. Barr and C. Wells, *Toposes, Triples and Theories*, Springer-Verlag, New York-Berlin-Heilderberg-Tokyo, 1985.

[7] A. Bauer, Five stages of accepting constructive mathematics, *Bull. Amer. Math. Soc.* **54-3** (2017) 481–498.

[8] L. Belair, *Calcul Infinitesimal en Géométrie Différentielle Synthétique*, M.Sc. Thesis, Université de Montréal, 1981.

[9] M. Berger and B. Gostiaux, *Differential Geometry: Manifolds, Curves, and Surfaces*, Springer-Verlag, Berlin-Heidelberg-New York, 1988.

[10] F. Bergeron, Objets infinitésimalment linéaires dans un modele bien adapté de la G.D.S., in: G. E. Reyes (ed.), *Géométrie Différentielle Synthétique*, Rapports de Recherche, DMS 80-12, Université de Montréal, 1980.

[11] J. M. Boardman, Singularities of Differentiable Maps, *Pub. Math. I. H. E. S.* **33** (1967) 21–57.

[12] F. Borceux, *Handbook of Categorical Algebra* **3**, Cambridge University Press, 1994.

[13] Th. Bröcker, *Differentiable Germs and Catastrophes*, Cambridge University Press, 1982.

[14] O. Bruno, Logical opens of exponential objects, *Cahiers de Top. et Géom. Diff. Catégoriques* **26** (1985) 311–323.

[15] O. Bruno, Vector fields on R^R in well adapted models of synthetic differential geometry, *J. Pure Appl. Algebra* **45** (1987) 1–14.

[16] M. C. Bunge, *Categories of Set-Valued Functors*, Ph.D. Thesis, University of Pennsylvania, 1966.

[17] M. C. Bunge, Relative functor categories and categories of algebras, *J. Algebra* **11** (1969) 64–101.

[18] M. C. Bunge, Internal presheaf toposes, *Cahiers de Top. et Géom. Diff. Catégoriques* **18-3** (1977) 291–330.

[19] M. C. Bunge, Sheaves and prime model extensions, *J. Algebra* **68** (1981) 79–96.

[20] M. C. Bunge, Synthetic aspects of C^∞-mappings, *J. Pure Appl. Algebra* **28** (1983) 41–63.

[21] M. C. Bunge, Toposes in logic and logic in toposes, *Topoi* **3** (1984) 13–22.

[22] M. C. Bunge, On a synthetic proof of the Ambrose-Palais-Singer theorem for infinitesimally linear spaces, *Cahiers de Top. et Géom. Diff. Catégoriques* **28-2** (1987) 127–142.

[23] M. C. Bunge, Cosheaves and distributions on toposes, *Algebra Universalis* **34** (1995) 469–484.

[24] M. C. Bunge and E. J. Dubuc, Archimedian local C^∞-rings and models of Synthetic Differential Geometry, *Cahiers de Top. et Géom. Diff. Catégoriques* **27-3** (1986) 3–22.

[25] M. C. Bunge and E. J. Dubuc, Local concepts in SDG and germ representability, in: D. Kueker et al. (eds) *Mathematical Logic and Theoretical Computer Science*, Marcel Dekker, New York and Basel, 1987, 93–159.

[26] M. C. Bunge and F. Gago Couso, Synthetic aspects of C^∞-mappings II : Mather's theorem for infinitesimally represented germs, *J. Pure Appl. Algebra* **55** (1988) 213–250.

[27] M. C. Bunge and M. Heggie, Synthetic calculus of variations, *Contemporary Mathematics* **30** (1984) 30–62.

[28] M. C. Bunge and P. Sawyer, On connections, geodesics and sprays in synthetic differential geometry, *Cahiers de Top. et Géom. Diff. Catégoriques* **25** (1984) 221–258.

[29] K. T. Chen, Iterated path integrals, *Bull. Amer. Math. Soc.* **83** (1977) 831–879.

[30] K. T. Chen, The Euler operator, *Archive for Rational Mechanics and Analysis* **75** (1981) 175–191.

[31] E. J. Dubuc, Adjoint triangles, in: *Reports of the Midwest Category Seminar II*, *Lect. Notes in Math.* **69** (1968) 69–91.

[32] E. J. Dubuc, Sur les modèles de la géométrie differentielle synthétique, *Cahiers de Top. et Géom. Diff. Catégoriques* **20** (1979) 231–279.

[33] E. J. Dubuc, Schémas C^∞, in [100], (1981) Exposé 3, pp 16–41.

[34] E. J. Dubuc, C^∞-schemes, *American J. Math.* **103–104** (1981) 683–690.

[35] E. J. Dubuc, Germ representability and local integration of vector fields in a well adapted model of SDG, *J. Pure Appl. Algebra* **64** (1990) 131–144.

[36] E. J. Dubuc and J. Penon, Objects compacts dans les topos, *J. Austral. Math. Soc.* **A-40** (1986) 203–217.

[37] E. J. Dubuc and G. E. Reyes, Subtoposes of the ring classifier, in: A. Kock (ed.), *Topos Theoretic Methods in Geometry*, *Various Publications Series* **30**, Matematisk Institut, Aarhus Universitet, 1979, 101–122.

[38] E. J. Dubuc and G. Taubin, Analytic rings, *Cahiers de Top. et Géom. Diff. Catégoriques* **24** (1983) 225–265.

[39] M. Dummet, *Elements of Intuitionism*, Clarendon Press, Oxford, 1977.

[40] C. Ehresmann, Les prolongements d'une variété différentiable I, II, III, *C. R. Acad. Sc. Paris* (1951), Reprinted in *Charles Ehresmann : Oeuvres Complêtes et Commentées. Part I*, Amiens, 1984.

[41] M. Fourman and M. Hyland, Sheaf models for analysis, in: *Applications of Sheaves, Lect. Notes in Math.* **753** (1979) 280–301.

[42] P. J. Freyd, Intrinsic Differential Geometry, Unpublished lecture, Philadelphia, 1983.

[43] A. Frölicher, Applications lisses entre espaces et variétés de Frechet, *C. R. Acad. Sci. Paris Ser. I Math* **293** (1981) 125–127.

[44] F. Gago Couso, *Internal Weak Opens, Internal Stability and Morse Theory for Synthetic Germs*, Ph.D. Thesis, McGill University, 1988.

[45] F. Gago Couso, Morse Germs in SDG, in: *Categorical Algebra and its Applications, Lect. Notes in Math.* **1348** (1988) 125–129.

[46] F. Gago Couso, Singularités dans la Géométrie Différentielle Synthétique, *Bull. Soc. Math. Belgique* **41** (1989) 279–287.

[47] M. Golubitsky and V. Guillemin, *Stable Mappings and their Singularities*, Springer-Verlag, New York-Heidelberg-Berlin, 1973.

[48] M. Goretsky and R. MacPherson, *Stratified Morse Theory*, Springer-Verlag, New York-Heidelberg-Berlin, 1987.

[49] I. M. Guelfand and S. V. Fomin, *Calculus of Variations*, Prentice Hall, Inc., London, 1963.

[50] R. Grayson, Concepts of general topology in constructive mathematics and sheaves, *Annals of Math, Logic.* **20** (1981) 1–41.

[51] V. Guillemin and A. Pollack, *Differential Topology*, Prentice Hall Inc, London, 1974.

[52] M. Heggie, *An Approach to Synthetic Variational Theory*, M.Sci. Thesis, McGill University, 1982.

[53] A. Heyting, *Intuitionism. An Introduction,* North Holland, 1956.

[54] M. W. Hirsch, *Differential Topology*, Graduate Texts in Mathematics **33**, Springer-Verlag, 1976.

[55] P. T. Johnstone, *Topos Theory*, Academic Press, Inc., London, 1977.

[56] P. T. Johnstone, *Sketches of an Elephant. A Topos Theory Companion, Volumes 1 and 2*, Oxford Science Publications, Oxford Logic Guides **43 44**, Clarendon Press, Oxford 2002.

[57] A. Joyal, Les théorèmes de Chevalley-Tarski el remarques sur l'Algébre constructive, *Cahiers de Top. et Géom. Diff. Catégoriques* **16** (1975) 256–258.

[58] S. C. Kleene, *Introduction to Metamathematics*, North Holland, 1962.

[59] A. Kock, Universal projective geometry via topos theory, *J. Pure Appl. Algebra* **9** (1976) 1–24.

[60] A. Kock, Formal manifolds and synthetic theory of jet bundles, *Cahiers de Top. et Géom. Diff. Catégoriques* **21** (1980) 227–246.

[61] A. Kock, *Synthetic Differential Geometry* (First and Second Editions), Cambridge University Press, 1981 and 2006.

[62] A. Kock, Synthetic characterization of reduced algebras, *J. Pure Appl. Algebra* **36** (1985) 273–279.

[63] A. Kock and R. Lavendhomme, Strong infinitesimal linearity, with applications to strong difference and affine connections, *Cahiers de Top. et Géom. Diff. Catégoriques* **25-3** (1984) 311–324.

[64] A. Kock and G. E. Reyes, Manifolds in formal differential geometry, in: *Applications of Sheaves. Proceedings, Durham 1977, Lect. Notes in Math.* **753** (1979) 514–533.

[65] A. Kock and G. E. Reyes, Connections in Formal Differential Geometry, in: A. Kock (ed.), *Topos Theoretic Methods in Geometry, Various Publications Series* **30**, Matematisk Institut, Aarhus Universitet, 1979, 158–195.

[66] A. Kock and G. E. Reyes, Aspects of fractional exponent functors, *Theory and Applications of Categories* **5** (1999) 251–265.

[67] S. A. Kripke, Semantical analysis of intuitionistic logic, in: J.N. Crossley and M. Dummett (eds.), *Formal Systems and Recursive Functions*, North-Holland, Amsterdam, 1965, 92–130.

[68] S. Lang, *Introduction to Differentiable Manifolds*, Interscience Publishers, New York London Sydney, 1967.

[69] R. Lavendhomme, *Basic Concepts of Synthetic Differential Geometry*, Kluwer Academic Publishers, 1996.

[70] F. W. Lawvere, Functorial semantics of algebraic theories, *Proc. Nat. Acad. Sci. U.S.A.* **50** (1963) 869–872.

[71] F. W. Lawvere, An elementary theory of the category of sets, *Proc. Nat. Acad. Sci. U.S.A.* **52** (1964) 1506–1511.

[72] F. W. Lawvere, Categorical Dynamics, Talk at the University of Chicago, May 1967, in: A. Kock (ed.), *Topos Theoretic Methods in Geometry, Various Publications Series* **30**, Matematisk Institut, Aarhus Universitet, 1979, 1–28.

[73] F. W. Lawvere, Quantifiers and sheaves, *Actes du Congrès International des Mathématiciens,* Nice 1970, 329–334.

[74] F. W. Lawvere, Towards the description in the smooth topos of the dynamically possible notions and deformations in a continuous body, *Cahiers de Top. et Géom. Diff. Catégoriques* **21** (1980) 377–392.

[75] F. W. Lawvere, Tiny objects, Unpublished talk at Aarhus University, June 1983.

[76] F. W. Lawvere, Outline of Synthetic Differential Geometry, Unpublished 1988. Available from http://www.acsu.buffalo.edu/∼wlawvere/SDG_Outline.pdf. Accessed 8 March 2017.

[77] F. W. Lawvere, Comments on the development of topos theory, in: J. P. Pier (ed.), *Development of Mathematics 1950-2000*, Birkhäuser, Basel-Boston-Berlin, 2000, 715–734.

[78] F. W. Lawvere, Categorical algebra for continuum microphysics, *J. Pure Appl. Algebra* **175** (2002) 267–287.

[79] C. McLarty, *Elementary Categories, Elementary Toposes*, *Oxford Logic Guides* **21**, Oxford University Press, 1995.

[80] M. Makkai and G. E. Reyes, *First-order Categorical Logic*, *Lect. Notes in Math.* **611**, Springer-Verlag, 1977.

[81] B. Malgrange, *Ideals of Differentiable Functions*, Oxford University Press, 1966.

[82] J. Martinet, *Singularities of Smooth Functions and Maps*, Cambridge University Press, Cambridge, 1982.

[83] J. N. Mather, Stability of C^∞-mappings I : The division theorem, *Annals of Mathematics* **87-1** (1969) 89–104.
[84] J. N. Mather, Stability of C^∞-mappings II : Infinitesimal stability implies stability, *Annals of Mathematics* **89-2** (1969) 254–291.
[85] J. N. Mather, Stability of C^∞-mappings III : Finitely determined map germs, *Pub. Sci. I.H.E.S.* **35** (1969) 127–156.
[86] S. Mac Lane, *Categories for the Working Mathematician* (Second edition), Springer-Verlag, Berlin-Heidelberg-New York, 1997.
[87] J. M. Milnor, *Morse Theory*, Annals of Mathematical Studies **51**, Princeton University Press, Princeton, N.J., 1969.
[88] J. M. Milnor and J. D. Stasheff, *Characteristic Classes*, Annals of Mathematical Studies **76**, Princeton University Press, Princeton, N.J., 1974.
[89] I. Moerdijk and G. E. Reyes, Smooth spaces versus continuous spaces in models for Synthetic Differential Geometry, Report **83-02**, University of Amsterdam, 1983.
[90] I. Moerdijk and G. E. Reyes, *Models for Smooth Infinitesimal Analysis*, Springer-Verlag, 1991.
[91] J. R. Munkres, *Topology from a Differentiable Viewpoint*, University of Virginia Press, 1966.
[92] R. Paré, Colimits in topoi, *Bull. Amer. Math. Soc.* **80** (1974), no. 3, 556–561.
[93] L. N. Patterson, Connections and prolongations, *Canad. J. Math.* **XXVII-4** (1975) 766–791.
[94] J. Penon, Infinitésimaux et intuitionisme, *Cahiers de Top. et Géom. Diff. Catégoriques* **22** (1980) 67–72.
[95] J. Penon, Le théoreme d'inversion locale en géométrie algébrique, Lecture given at Louvain-la-Neuve, *Journées de faisceaux et logique*, May 1982.
[96] J. Penon, *De l'Infinitésimal au Local*, Thése de Doctorat d'Etât, Université Paris VII, 1985, *Diagrammes*, Paris VII, 1985.
[97] V. Poénaru, *Analyse Différentielle*, Lect. Notes in Math. **371**, Springer-Verlag, 1970.
[98] W. A. Poor, *Differential Geometric Structures*, McGraw Hill, 1981.
[99] N. Van Quê and G. E. Reyes, Smooth functors and synthetic calculus, in: A. S. Troelstra and D. van Dalen (eds.), *The L.E.J. Brower Centenary Symposium*, North-Holland, Amsterdam-New York-Oxford, 1982, 377–396.
[100] G. E. Reyes, (ed.), *Géométrie Différentielle Synthétique, Rapports de Recherche du Département des Mathématiques et Statistiques* **80-11, 80–12**, Université de Montréal, 1980.
[101] G. E. Reyes, A remark of M. Makkai on Postulate WA2 [25, 35]. Private communication, July 14, 2009.
[102] A. Robinson, *Non-Standard Analysis*, North-Holland, 1966.
[103] A. M. San Luis Fernández, *Estabilidad Transversal de Gérmenes Representables Infinitesimalmente*, Ph.D. Thesis, Universidad de Santiago de Compostela, 1999.
[104] A. Sard, The measure of critical values of differentiable maps, *Bull. Amer. Math. Soc.* **48** (1942) 883–890.
[105] R. Sikorski, *Boolean Algebras*, Springer-Verlag, Berlin, 1964.
[106] R. Thom, Une lemme sur les applications différentiables, *Bol. Soc. Mat. Mex.* **2-1** (1956) 59–71.

[107] R. Thom, *Stabilité Structurelle et Morphogénèse: Essai d'une Théorie Générale des Modèles*, Benjamin-Cummings, 1972.

[108] R. Thom and H. L. Levine, Singularities of Differentiable Mappings I, Bonn 1959, Reprinted in: C. T. C. Wall (ed.), *Proceedings of Liverpool Singularities Symposium*, *Lect. Notes in Math.* **192**, Springer-Verlag, 1972.

[109] A. S. Troelstra, *Principles of Intuitionism*, *Lect. Notes in Math.* **95**, Springer-Verlag, 1969.

[110] G. Wassermann, *Stability of Unfoldings*, *Lect. Notes in Math.* **393**, Springer-Verlag, 1970.

[111] A. Weil, Théorie des points proches sur les variétés differentiables, *Colloq. Top. et Géom. Diff., Strasbourg*, (1963) 111–117.

[112] S. Willard, *General Topology*, Addison-Wesley, 1968.

[113] D. Yetter, On right adjoints to exponential functors, *J. Pure Appl. Algebra* **45** (1987) 287–304. Corrections to "On right adjoints to exponential functors", *J. Pure Appl. Algebra* **58** (1989) 103–105.

Index

Printed in the United States
By Bookmasters